International Conference on Marine Design 2020

Cadiz, Spain
15 – 16 January 2020

ISBN: 978-1-7138-0870-1

Printed from e-media with permission by:

Curran Associates, Inc.
57 Morehouse Lane
Red Hook, NY 12571

Some format issues inherent in the e-media version may also appear in this print version.

The Royal Institution of Naval Architects
8-9 Northumberland Street
London, WC2N 5DA
United Kingdom

Phone: 020 7235 4622
Fax: 020 7259 5912

publications@rina.org.uk

Additional copies of this publication are available from:

Curran Associates, Inc.
57 Morehouse Lane
Red Hook, NY 12571 USA
Phone: 845-758-0400
Fax: 845-758-2633
Email: curran@proceedings.com
Web: www.proceedings.com

International Conference
MARINE DESIGN 2020

15th-16th January 2020, Cádiz, Spain

PAPERS

THE ROYAL INSTITUTION OF NAVAL ARCHITECTS
8-9 Northumberland Street, London, WC2N 5DA, Telephone: +44 (0)20 7235 4622

The Royal Institution of
Naval Architects

INTERNATIONAL CONFERENCE

Marine Design 2020

15th -16th January 2020, Cádiz, Spain

CONTENTS

DESIGNING WATER TAXIS

A. Nazarov, Albatross Marine Design, Thailand

SUMMARY

Nowadays, intercity water transport in Bangkok, Sydney, or Dubai is not only an important element of mass passenger transit, but also a sightseeing attraction and integral part of city landscape. This paper presents a vision of a new generation of the smallest passenger watercraft, water taxis, designed for modern urban environments. These boats are seen as highly efficient light but safe and comfortable craft, with modern technologies of electrical propulsion and use of solar energy. Other concerns are low wave-making and low noise. The ergonomics of water taxis is studied in terms of feasible dimensions and possible compromise with requirements for bigger ships. Typical interior layouts are discussed and compared. Seating arrangements, boarding options, luggage storage areas for comfort and safety are analyzed. Special attention is given to accessibility for passengers with disabilities or mobility restrictions. New designs of water taxis on catamaran schemes are presented, based on extensive design experience of AMD team in field of small passenger catamarans. Specifics of safety assessment related to exterior styling and layout are reviewed, and recommendations to applicable rules and standards are provided.

Figure 1: F100 - 10m water taxi for designed by AMD. This design became the Gold Winner at European Product Design Award 2019.

1. INTRODUCTION

Passenger transportation by boats is an important part of daily life in many parts of the world. From simple traditional craft to modern high-speed ferries, these boats carry both urban and rural residents, tourists or personnel. Most of the craft used for such transport can be categorized as small (i.e. with hull length L_Hδ24m) and operate between islands, in coastal areas, rivers and canals.

Such small passenger boats often inherit local functional and aesthetic features and are an integral part of the waterfront landscape. On the other side, the image factor is important: remarkable, elegant, and comfortable boats are eye-catching and represent the technological level of a country.

A large number of such projects of small passenger craft has been developed by Albatross Marine Design during recent years, including some award-winning concepts (fig.1). A A A number of studies has been published [2-4,6], covering various aspects of design, operational costs, and safety. The present paper represents a summary of this experience and is intended for transport designers and naval architects. Here we do not touch on such specific cases as air-cushioned vessels, hydrofoils, ground-effect craft as they are not yet in AMD's portfolio.

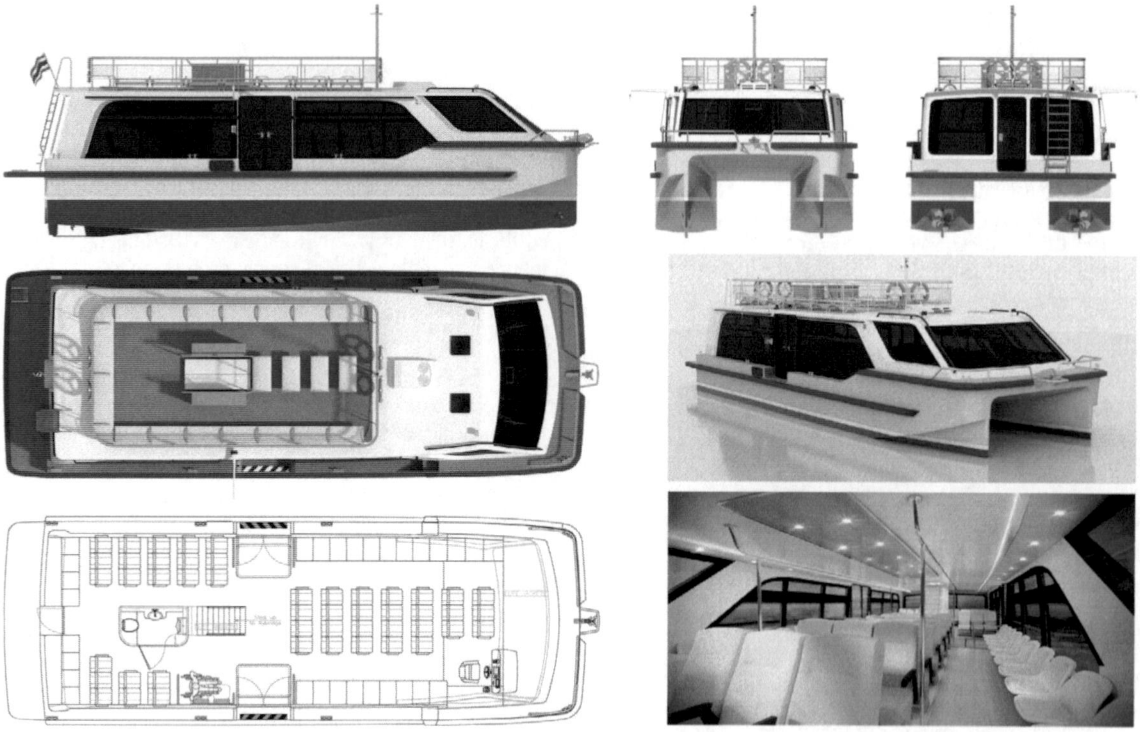

Figure 2: F150EC1 15.7m shuttle boat for designed by AMD for company operating cruising liners.

2. TYPES OF CRAFT

When talking about passenger craft, definitions are important and can be very confusing. Say, for a typical recreational charter small craft under ISO8666 all persons on board are 'crew', whether they are involved in actual handing of craft or not. Typical types of small passenger craft according to the tasks performed are:

- Water taxis (fig.1, 6) – passenger craft used for hire on custom routes with a small number of passengers on board;
- 'Transporters' – primarily used for transportation of passengers from A to B (fig.5,8);
- Shuttles (fig.2,3) - used by hotels, waterfront condos, and cruise operators to transport their clients short distances;
- Tour boats (fig.4,7) – used for transportation associated with other activities such as sightseeing, excursions, island trekking, watersports, etc.;
- Purpose-built boats, such as glass bottom craft, dive boats, etc.

In terms of capacity, small passenger craft can be subdivided into:

- up to 12 passengers, which are not formally considered as passenger craft, but are actually used as such;
- between 36 and 50 passengers where in many jurisdictions simplified requirements are used;
- over 50 passengers – often treated as large craft in terms of safety requirements.

Note that from the viewpoint of tour operators these categories are related to a minivan (up to 12) or bus (below 50) passengers.

3. SAFETY CONSIDERATIONS

Small passenger craft is a kind of grey area with no unified requirements, so rules for bigger craft are often applied by aboriginal classification societies causing a large number of compliance problems. At the other end of the spectrum, there are jurisdictions where no basic safety standards exist at all. In reality, safety requirements are often dependent on local authorities and should be negotiated on a case-by-case basis.

Small passenger craft are operated predominantly in coastal or inland waters, so where accidents occur rescue can be provided promptly. Most small craft are also fast enough to reach shelter if unfavourable weather conditions are expected. Thus, an equivalent safety approach should be applied, with real case scenarios of every accident considered.

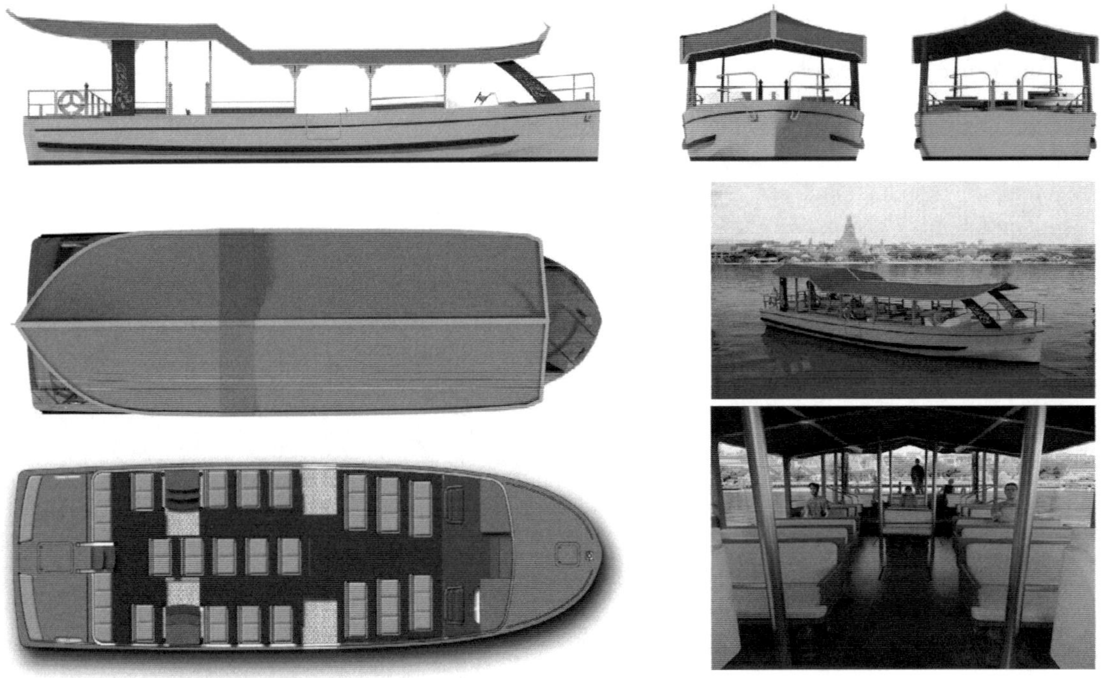

Figure 3: MOH15 15m shuttle boat for luxury hotel in Bangkok

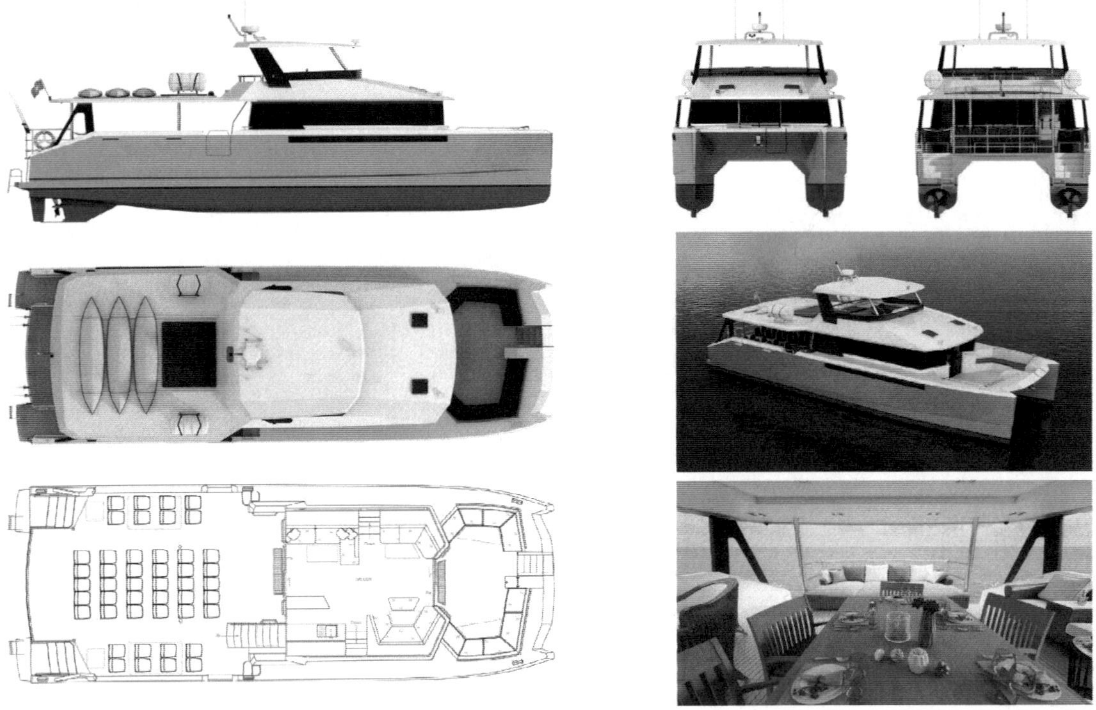

Figure 4: RT17 17m tour boat for tropics. The boat is designed for different types of activities - from transport and day trips, to private parties and dining cruises.

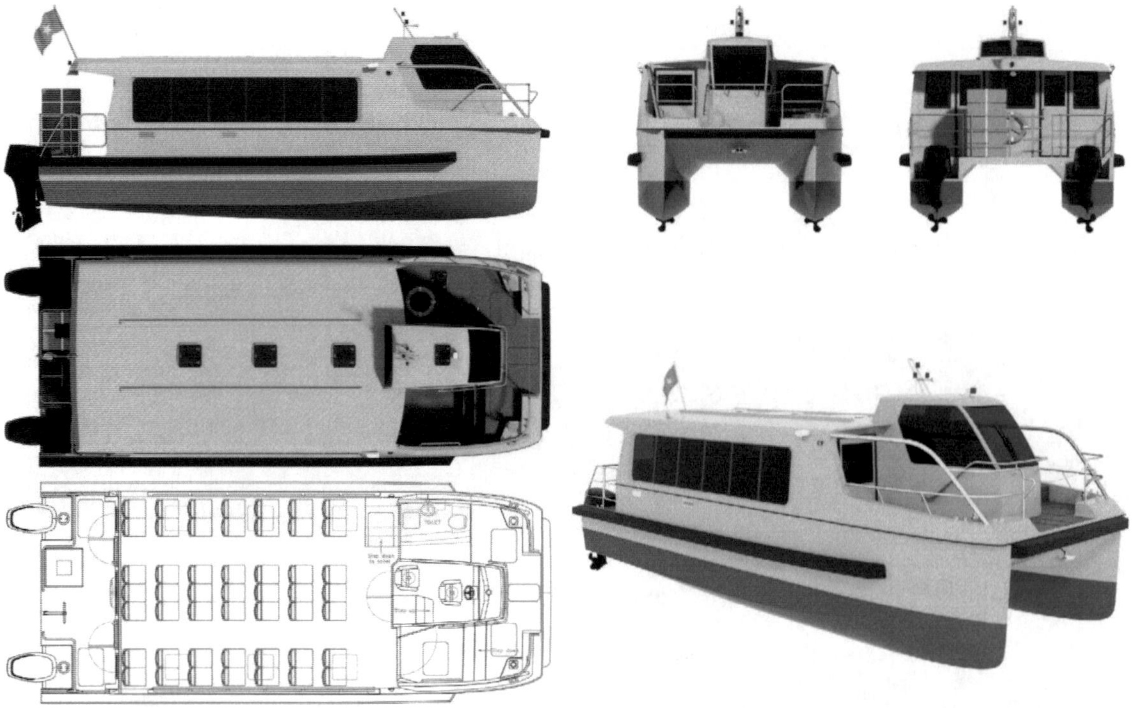

Figure 5: H12 12m passenger boat for Vietnam. The boat is designed for both bow loading and side loading, hull material - composite.

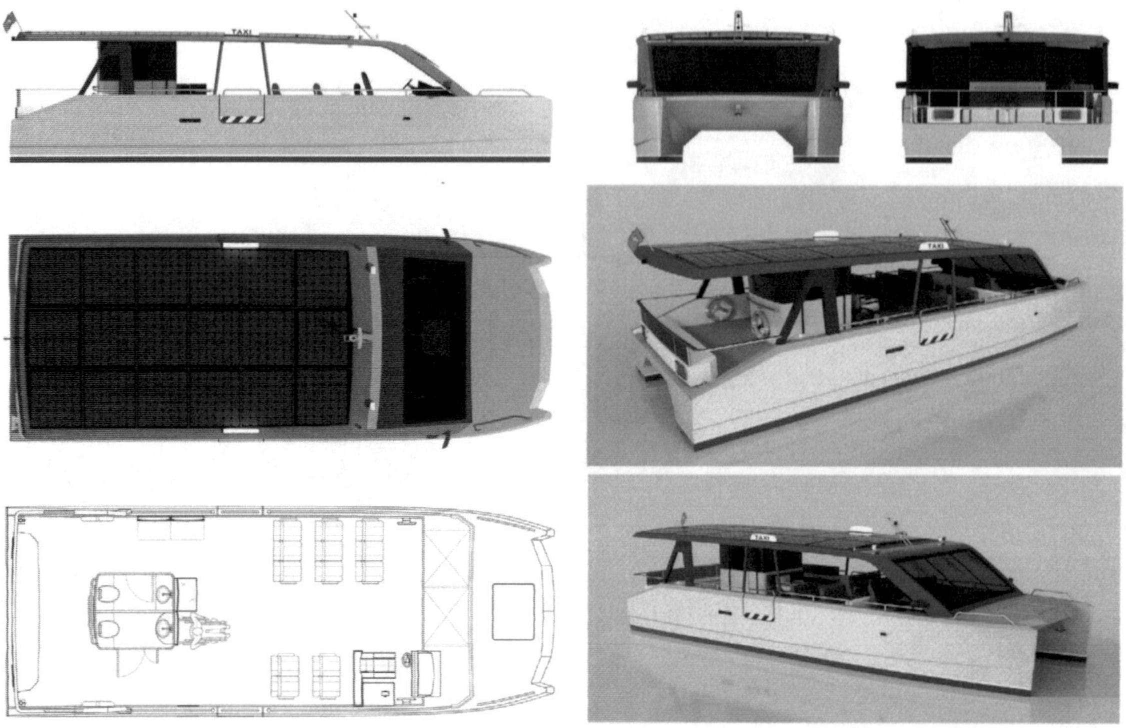

Figure 6: RT12 12m water taxi boat. This boat can be used for private trips or for parties on board, in tropical destinations.

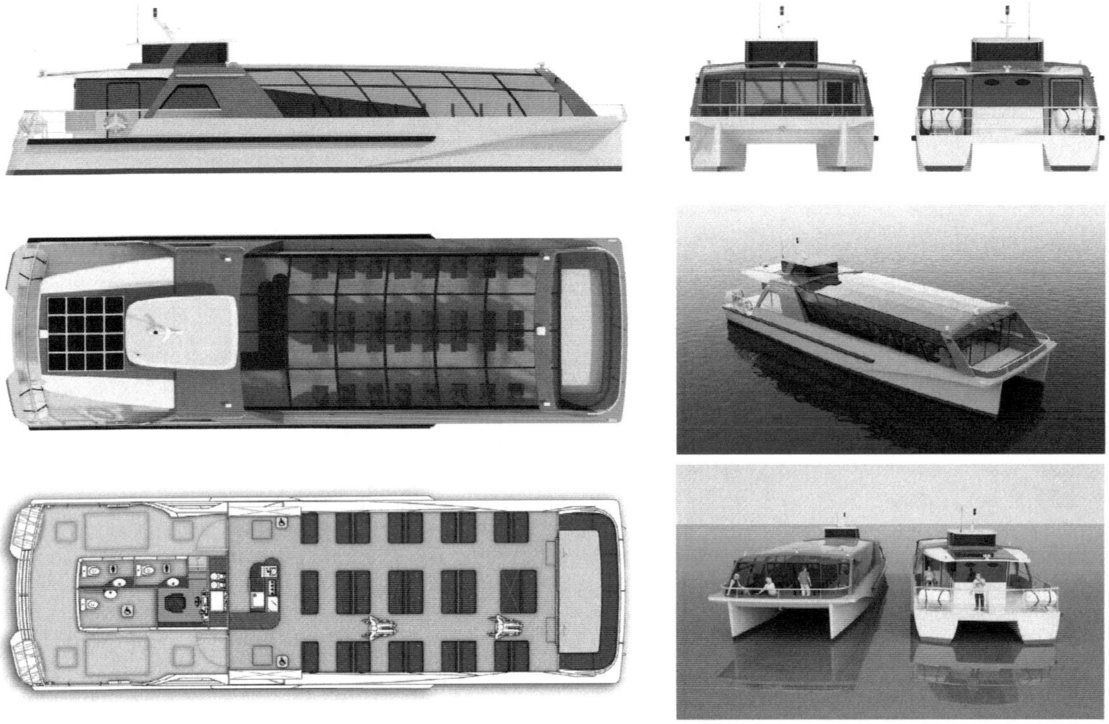

Figure 7: MWT 20m tour boat for Moscow river.

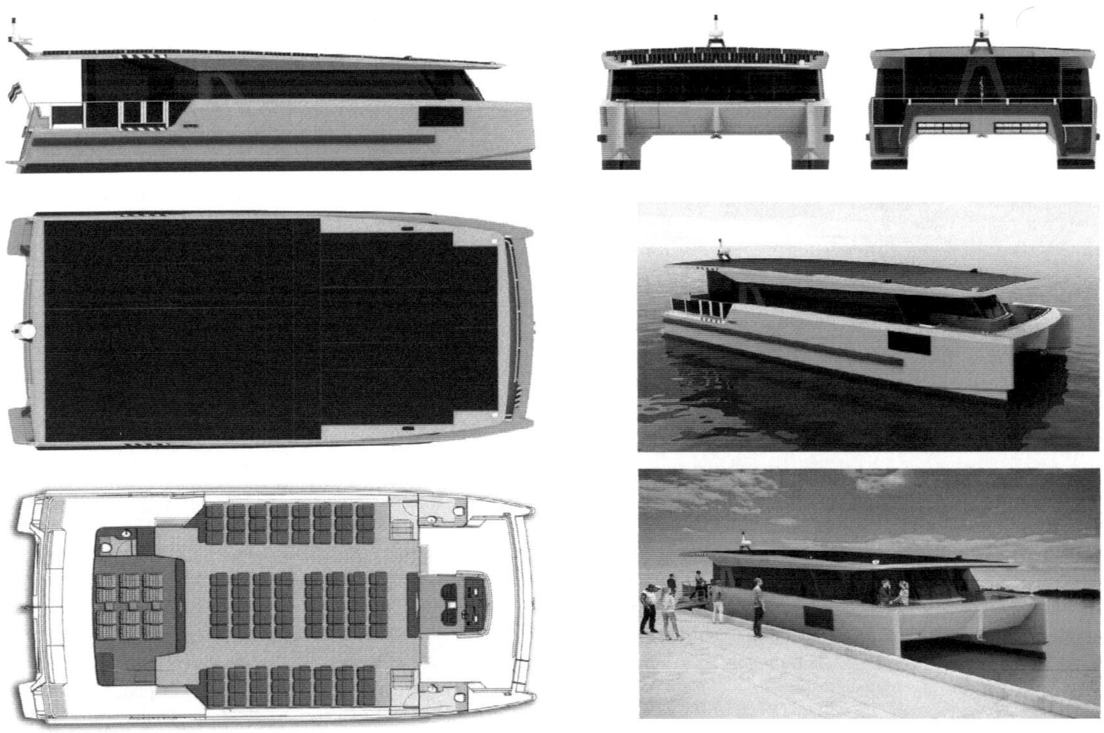

Figure 8: AMD20 20m solar-powered passenger boat.

As an example, due to the limited number of passengers evacuation time from small passenger craft is quite short. In case of fire, reduction of structural fire protection time from regular 60 to 15 minutes is possible.

In [4] featured are the most popular rules and standards applicable to small passenger craft. As an addition, Australian NSCV [5] is mentioned as a good example of a reasonable set of rules.

In AMD's practice, in-house safety matrix, based on applicable rules, experience and common sense are used. The part of this safety matrix affecting general design and styling is presented in Table 1. This table is the basis for initial design and further certification negotiations with authorities.

However, beyond the rules and regulations, in reality most small passenger boats are often overloaded in service. This happens due to poor loading control by the crew, a policy of the operating company, or the effect of being the last shuttle of the day. Thus, it is important for the designer and naval architect to foresee the potential problems by providing overload-proof designs. This relates to management of areas accessible to passengers and restricting areas where crowding and overloading can occur. Frankly speaking, at this level safety mostly depends on the designer who deals with real-world safety, rather than on the authorities who mostly care about paper safety only. As our recommendation, an overload factor of at least 20 to 30% of passenger capacity should be accounted for in any design when calculating freeboard and stability. Technically this means that catamarans are often the best type of passenger craft providing safety from possible overloading.

4. CRAFT SIZING AND LAYOUT

Sizing of small passenger craft depends on many factors, the most important of which are:

- Acceleration/motion comfort;
- Saloon layout and type of activity on board;
- Local restrictions, such as passenger or length limits for easy registration (see fig.5);
- Docking and transport restrictions.

Quite often the client requests a craft size which is too small for the desired operation. One should bear in mind that the smaller the craft, the less comfortable it is in terms of motions and acceleration causing seasickness. The size of craft should be analysed in terms of displacement - speed - operational sea state, at early design stages. Passenger seats should be preferably located in the zones of best motion comfort of the craft.

Statistics of small high-speed craft accidents provided in [2,6] indicates that the craft's arrangement is critical for safety. On fast boats with sideways seating up to 70% of passengers experience injuries following collision accidents. Though formally small passenger craft are not subject to IMO HSC Code [1], some good design practices reflected in the Code are useful when planning craft's layout. In such cases no seating at the bow area as well as direction of seating should be considered (Table 1).

Small passenger craft (and its dimensioning) are designed around passenger saloon and seating arrangement. Dimensions of saloon are discrete in terms of seat unit inclusive of passageways and service areas. Deck area required for saloon is 0.5 to 0.9 m^2 per one passenger, where lower numbers are typical for shorter trips and/or smaller craft (see fig.2-8). Detailed diagrams on sizing of saloon and passenger seats are presented in [4].

5. HOTEL SHUTTLE BOAT 'THAI STYLE'

The client for this project is a luxury hotel operator in Bangkok. Being located in the heart of a modern city, the operator is facing issues with road traffic. In such a situation, transportation by boats is an attractive and fast option for luxury hotels. For this project styling, combination of cultural background, existing Oriental style of hotel, and site landscape were considered.

The boat has dual purposes: daytime shuttle, or dining cruise in the evening. Two boarding zones are provided for both high and low types of jetties, available on Chaphraya river.

The traditional boats used on the river currently are quite heavy and built of timber. Seasonal maintenance cost is high, and with limitations on the use of wood the future of these craft is uncertain. Thus, AMD team was facing the challenge of keeping an authentic traditional look but of making improvements of performance, comfort, and safety. The boat's hull structure is designed in composite and the design in general is made in compliance with DNVGL0314 standard.

The most challenging task was the design of Thai-style decorations which should be outsourced to woodwork carving shop and attached to composite structure.

6. WATER SHUTTLE CATAMARAN

This project (fig.2) started from the requirement to design a boat for transportation of passengers from cruise liners to shore, at tropical destinations where no port infrastructure exists. Currently, such transportation is carried out by on-board lifeboats which is probably not the best solution in terms of comfort and accessibility for passengers with restrictions in mobility. The new craft should be a catamaran suitable for road transportation twice a year, to respond to seasonal changes of cruising destinations. From our survey of facility on board of cruising ships, and inputs from the operator, the following brief of design requirements was formulated:

- length 15m, beam 5.5m
- freeboard 0.9 to 1.0m at boarding station
- speed max/cruising 25/20 kts
- passenger capacity 120-150
- double deck
- material – aluminium
- operational sea state – up to 3
- suitable for disabled passengers

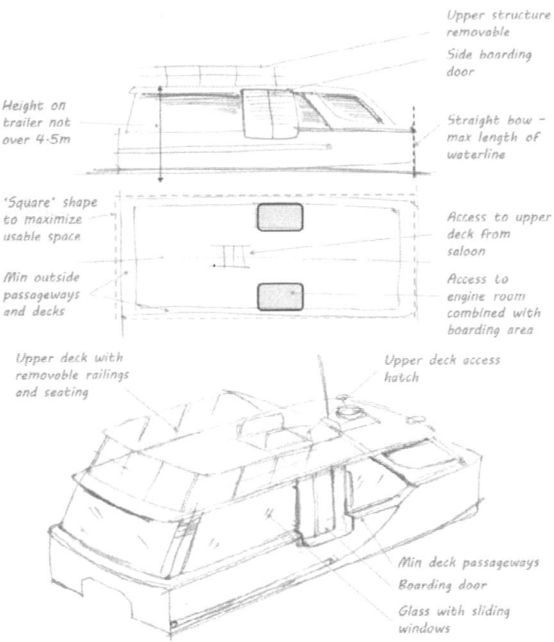

Figure 9: First sketches of 15m water shuttle: some ideas presented to the client during first meeting.

Based on the preliminary study, the following design features are essential (fig.9):

- Compact design with limited overall dimensions
- Maximum saloon area for maximum seating
- Height limit for transportation
- Upper deck railings removable
- Boarding area is shared for engine room access
- Nice panoramic view from saloon
- Attractive exterior – signature of tour operator
- Safety to international standards (as practically possible on a small boat)

Operational speed requirement was questioned by our design team. For short transport legs (below 15 minutes) speed is not important, however higher speeds will require larger engines and heavier hull structures, as well as a seating arrangement that has to comply with high-speed craft requirements. Operational speeds at hump of resistance should be avoided in any case. On such a heavy loaded craft with limited length recommended operational speeds should not exceed 10 to 12kts.

7. TOUR CATAMARAN FOR TROPICS

This 17m tour catamaran concept (fig.4) was designed at the demand of a boat manufacturer looking to enter the tourist boat market. The concept is based on 15m and 17m designs drawn by AMD in previous years; those boats proved to be attractive and efficient carrying tourists in South-East Asia and Oceania.

The new RT features few functional zones, which are available for passengers and parties on board. The forward cockpit is a great place for sunbathing and sightseeing once the craft is underway. Bow ramp is available for beaching. Flybridge features steering station and passenger seats. Space in cabin in mostly utilised for the bar and protected seating. Cabin windows are vertical to maximize space, to reduce direct sunlight, and for ease of forward door installation. Aft deck is spacious and can be converted into passenger seating area or party zone. Simple cabins and bathrooms are on the lower deck, accessible from saloon. The craft carries a lot of toys, such as kayaks and inflatables, and also recreational diving equipment. Hull material is a composite sandwich.

8. PASSENGER CATAMARAN FOR VIETNAM

This is an example of a boat (fig.5) designed for a regulatory niche: around 12m length limitation to use outboard engines and reduced crew. Outboard engines allow the reduction of the initial cost of the craft, however they are not optimum if operated at long distances [6]. Bow loading was requirement for operation in Na Chang.

The most difficult thing for this craft is the requirement for certification by Vietnam Register of Shipping. This is a tricky procedure where the craft is reviewed by surveyors who normally work with offshore drilling rigs and cargo ships. Thus, decent weight and space margins should be foreseen by the designer to allow for absolutely necessary items such as main and backup generators and fire pumps, which are likely to be required by the surveyors.

9. 12M WATER TAXI CATAMARAN

This catamaran (fig.6) was designed as a development of an award-winning 10m design shown on fig.1. Electrical propulsion requires improved propulsion efficiency, thus longer and lighter hulls were designed for the new 12m version.

RT12 is our vision of a water taxi for modern urban environments, highly efficient, light but safe and comfortable craft, with modern technologies of electrical propulsion and use of solar energy. Other concerns were low

Table 1 - Sample of safety martix of 'good practice requirements' developed and used 'in house' by AMD (only items relared to general architecture and layout are shown)

№	Requirement	Recreational or commercial up to 12 passengers	13-50 passengers	over 50 passengers
1	Certification	CE-certificate or equivalent	CE-certificate or equivalent	The craft should be designed to comply with basic safety requirements listed below, or class of passenger vessel
2	Intact stability – compliance to ISO12217 including offset load test, or compliance to IMO/HSC Code Annex 7 or 8.	Compliance to ISO12217 including offset load test, or compliance to IMO/HSC Code Annex 7 or 8, or equivalent.	Compliance to ISO12217 including offset load test, or compliance to IMO/HSC Code Annex 7 or 8, or equivalent.	Compliance to IMO/HSC Code Annex 7 or 8, or equivalent standard. Stability book to be provided on board
3	Passengers should be seated in forward or aft facing seats with armrests or seat belts. No passenger seating at forward part of craft.	Not required	Recommended, if cruising speed of craft exceeds 20kts	Required, if cruising speed of craft exceeds 20kts. Passenger accommodation to comply IMO HSC Code.
4	For boats over 15m in length, watertight collision bulkhead located at 5-10% of length from forward end of LWL. This requirement does not apply to boats with built-in buoyancy elements.	Recommended	Required	Required
5	For decked boats over 15m in length – watertight engine room bulkhead(s)	Recommended	Required	Required
6	Damaged floatation and stability in case of flooding of any one compartment. This requirement does not apply to boats with built-in buoyancy elements.	Not required	Required	Required
7	Width of escape routes measured to ISO9094 should not be less than	500mm	600mm	900mm (to IMO standards)
8	Emergency exists labelling	per ISO standards	ISO standard or IMO symbols	IMO symbols
9	Structural fire protection in engine room (for closed engine rooms only)	not required	B15 or similar	A0 or similar on metal craft B30 or equivalent on non-metallic craft
10	Fire safety	ISO9094	ISO9094	ISO9094 or class rules
10	Liferafts or similar rescue equipment	depending on service area	minimum 1, capacity 100% of persons on board	minimum 2 liferafts, capacity: 100% of persons on board should be taken by all liferafts minus one.
11	Life jackets	Approved-type life jackets 110% of persons on board; plus 10% of children life jackets. For vessels engaged in tours with larger percentage of children additional number of children life jackets is recommended.		
12	Lifejackets racks labelling	visible	IMO symbols	IMO symbols
13	Public announcement system	not required	recommended	required, in all habitable spaces
14	Deck safety - railings	compliance to ISO 15085 Small craft - Man-overboard prevention and	minimum 750mm high	minimum 1000mm high

wave-making and low noise: on most of rivers and canals waves washing the banks represent big problem.

The catamaran offers transport for 12 to 20 passengers at comfortable speeds of up 8 to 15kts, with comfortable motions of the catamaran platform. The catamaran is highly manoeuvrable in tight areas due to the wide separation of propulsion units. Operating in choppy conditions of harbours/rivers with a lot of traffic this wide and high freeboard catamaran will remain dry and stable.

There are designated wheelchair zones, with adjacent seats for accompanying persons. Priority was given to lightness, roomy public spaces, natural airflow, practicality, and easy maintenance at affordable cost. The structure would be in composite sandwich with carbon reinforcements. However, safety features are not compromised: the boat complies to ISO Small Craft Standards for category C.

9. TOUR CATAMARAN FOR MOSCOW

This boat is intended for operation in rivers, and inland waterways, for tourism and intercity transport. The exterior is a floating glasshouse structure look, but is modern and highly functional. The saloon is covered by glass structure with adjustable tint. There are no side deck passageways where standing passengers would normally block the view of those seated in the saloon. Instead, for photography and sightseeing a bow balcony is used. The pilothouse is half-deck elevated, providing panoramic visibility yet is within 4.5m air draught (with folding mast) to pass under bridges. Bathrooms are accessible from the aft deck; this area is also used for boarding. The layout of the vessel is anticipates a few options: from intercity bus accommodation and sightseeing tours layouts, both for 60 passengers, to dining cruise, and special events options for 40 guests.

In terms of styling, the craft looks in harmony with modern cities, with a certain level of familiarity for the typical urban dweller. The boat is wheelchair friendly: loading of wheelchairs and their movement around the craft is accommodated.

The hull shape combines fine straight bow entries with maximized deck space and effective spray deflection. Powering is by a hybrid system with conventional diesel engines as an option. Material is aluminium for hulls, with composite/glass superstructure.

10. 20M SOLAR POWERED CATAMARAN

The boat presents a designers' vision of an eco-friendly water transport for modern urban environments. The new boat is designed around use of low emission and low noise electrical propulsion system. Full roof is maximized in size, and is used to locate 135m^2 of solar panels, capable of providing 30kW of power on tropical sun. Any shading of solar panels is unfavourable; thus the pilothouse is designed to be level with the main saloon.

The boat is intended for transport of 100-120 passengers on rivers or protected waters, with a cruising speed of 10kts, with low wave-making and comfortable motions of catamaran platform. Boarding of passengers is through side doors, and also features an optional bow ramp for bow loading or beach loading. The boat features main saloon with seats, and also an optional business class saloon at the aft deck; there are 2 or 3 bathrooms. Special thinking was given to comfort of passengers with restrictions in mobility.

Style of boat is minimalistic with all practical features, with priority given to lightness, roomy public spaces, natural airflow, and easy maintenance at affordable cost. However, safety features are not compromised. The catamaran is highly manoeuvrable in tight areas due to wide separation of propulsion units.

The design of hull shape and functional features is based on extensive design experience of AMD team in the field of passenger catamarans. Hull structure is seen in composite materials which allow for a lighter and stronger hull.

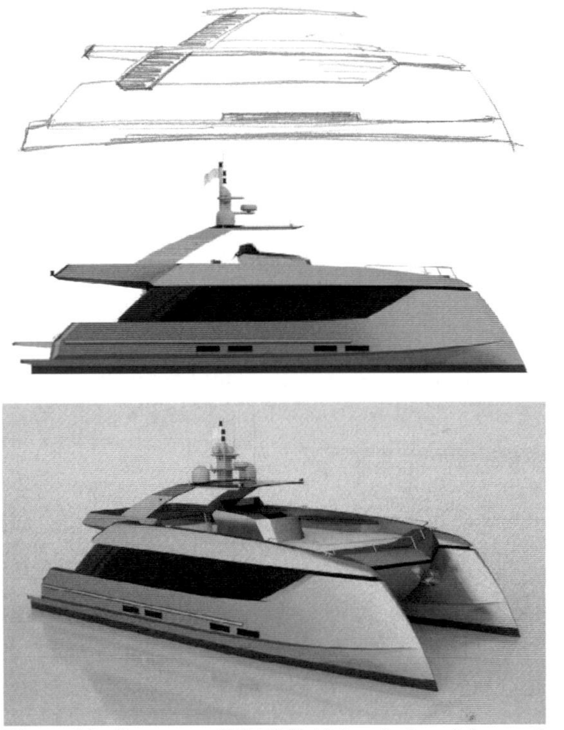

Figure 10: Catamaran TCAT60 18.2m designed for parties and group tours.

11. CONCLUSION

Design of small passenger craft is a creative task combining aesthetical and technical components with relation to economic considerations.

In many locations, operation of water taxis and small passenger craft is growing due to low operational costs and easy certification, shallow draft and berthing limitations, and the demand for custom tasks and destinations. For small businesses, entrance opportunities to the water transportation market with minimal initial investment is important. This is one more argument in favour of smaller boats. The variety of designs shown in the present paper are developed by AMD during 2018-2019; currently on the drawing board is party catamaran fig.10. This is a clear indication of the high demand for such boats, so great opportunities exist for designers in this market.

The present paper and case studies provided would be helpful for marine design students and industrial/transport designers willing to advance the development of efficient and safe passenger boats.

12. REFERENCES

1. International Code of Safety for High Speed Craft (2000 HSC Code) – IMO, 2008 Edition
2. Latorre R.G., Nazarov A. Functionality and Safety of Small Passenger Craft: Some Lessons// 5th Chesapeake Power Boat Symposium - Annapolis, USA, 2016. Nazarov etc. Small passenger catamarans
3. Nazarov A. Practical Small Craft Design: Combining Art with Science// The Transactions of The Royal Institution of Naval Architects, International Journal of Marine Design Vol. 154 Part C1 2012.
4. Nazarov A., Jabtanom A., Charatsidis N. Small Passenger Boats and Water Taxis: Aesthetic and Functional Aspects of Design //Marine Design Conference, RINA, 2014 - Coventry, UK. p.83-94.
5. NSCV Australian Transport Council - National Standard for Commercial Vessels. 2011.
6. Saenbunwang S., Sawatpanich S., Potgieter W., Nazarov A. Small Passenger Craft for South East Asia: Design Specifics and Case Studies// China International Boat Show & High Performance Marine Vessels Conference (HPMV 2015).

12. AUTHORS BIOGRAPHY

Dr. **Albert Nazarov** holds the degree in Naval Architecture from Sevastopol National Technical University (1996) and Ph.D. in Ship Dynamics from Odessa National Maritime Academy (2004), he is a Fellow of RINA, CEng, a member of SNAME, and a member of Designers Union of Russia. His current position is managing director at Albatross Marine Design (Thailand) with experience in construction, design and research of small and high-speed craft.

ESPOMAR PROJECT: NEW PASSENGER MARITIME TRANSPORT SYSTEM FOR THE GULF OF CÁDIZ. (PRELIMINARY RESULTS)

A. Querol, P. Piniella, P. Sánchez, J.C. Rasero, L.A. Fernández, S. Sotomayor, M.C. Pérez, D. Sales, University of Cádiz, Spain
A. Vargas, University of Huelva, Spain
F. Perna, University of Algarve, Portugal

SUMMARY

The ESPOmar project is a cooperation research network with the aim of designing and evaluating a new sustainable cross-border maritime transport system for the Gulf of Cadiz. The key objectives of the ESPOmar project are: to analyse the real demand of such a transport system; to study the nautical condition for navigation of the area; to define the capacity and specification of each proposed maritime route; and to optimize the design of specific ferries to compete with the established land transport system in terms of speed, safety, and comfort, reducing emissions and increasing economic viability. Economic viability and local economic impact were key tenets of the evaluation process. This paper presents the preliminary results and conclusions of the evaluation process for the proposed maritime routes examined.

1. INTRODUCTION

The ESPOmar Project, co-financed by the European Regional Development Fund (ERDF) through the INTERREG V-A Spain-Portugal (POCTEP) 2014-2020 program, is a co-operation network with the aim of Designing a Sustainable and Transboundary Transport System Maritime-Fluvial in the Gulf of Cádiz, in which the Spanish Universities of Cádiz (UCA) and Huelva (UHU), the Portuguese University of Algarve (UAlg), and the Public Ports Agency of Andalusia (APPA) have all participated.

The purpose of this project focuses on the area of the Gulf of Cádiz, which due to its physiognomy and the existence of both Doñana Natural Park and Guadalquivir River, makes direct land connections between nearby provinces much longer than possible maritime connections. Figure 1 shows how a maritime connection between Cadiz and Faro would have a reduction in distance of more than 150 kilometres and 105 kilometres between Cadiz and Huelva.

The most important and consideration to take into account in the development of the project is the Doñana Natural Park, along with the mouth of the Guadalquivir River. This is because both areas provide refuge and shelter for an immense variety of birds and fish, being considered the most important wetland in the entire European continent. Over the last seventy-five years the idea of building a direct road to connect Cadiz and Huelva has been discussed. Originally, that road was projected as a coastal road over the Guadalquivir River and through the National Park. Now that project has been completely withdrawn but some political groups are still planning a new road through the north exterior border of the National Park. Some civil and political organizations are really concerned about the environmental impact that this road could have from the point of view of protection of the natural space, as well as from the perspective of public expenditure. They would prefer to isolate the protected area completely and turn Doñana into a kind of zoo rather than a natural area. On the other hand, this measure, even from an economic point of view, could prevent the promotion of tourism in national parks.

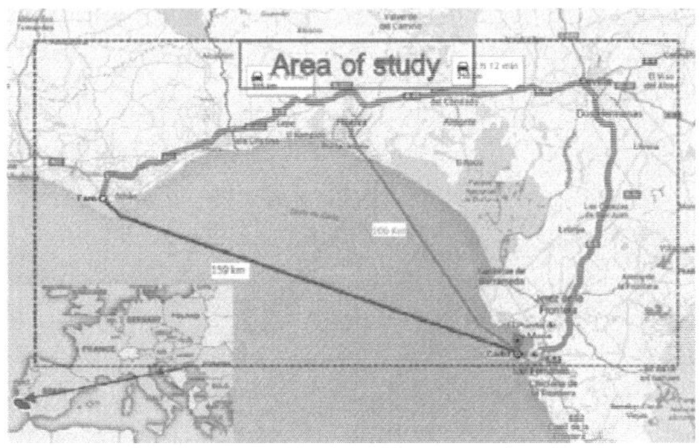

Figure 1. Gulf of Cádiz (area of study)

The existence of other motorways in service or in execution and environmental requirements discourage their construction, but that does not mean that both provinces continue to call for an improvement in connections. There are even political options that put the construction of these highways before environmental criteria. In addition, with this project, an attempt is made to provide a more sustainable alternative to the direct connection that is still non-existent between the bordering provinces of Cádiz and Huelva. Currently these provinces are linked by public transport (bus and rail), but not directly. In the case of the bus, in Seville it is necessary to transfer from one bus station to a different one where another transport company takes passenger from Seville to Huelva and vice versa.

Another of the objectives of the ESPOmar Project is the participation in the sustainable development of this area, improving connectivity, the tourist offer, the use of heritage resources, the management of natural resources, and the effort to reduce greenhouse gas emissions. In this way, the natural and cultural potential of the area would be exploited in a non-invasive way, so that this would contribute to a greater and better economic development of all this part of the Iberian Peninsula coast. To carry out this research a series of activities was established that will be fully developed in this paper:

- Demand study and analysis
- Design of transport lines
- Design of sustainable vessels
- Environment effect investigation
- Economic analysis
- Optimization

2. METHODOLOGY

The methodology used in the different activities that are being developed during the execution of this project is detailed below:

2.1 DEMAND STUDY AND ANALYSIS

The aim of this activity is to determine the possible demand for a new maritime transport system, information which is needed to correctly adjust the capacity of the lines and determine the economic and environmental viability of the transport system. The information included in this demand study are the following:

- Frequency and characteristics of displacements between the study populations (Cádiz, Huelva and Portuguese region of Algarve).
- Predisposition to travel by boat
- Importance of factors or conditions to travel by boat:
    - Travel time from the starting point to the departure port and from the arrival port to the destination
    - Weather conditions
    - Quantification of the maximum price to be paid
    - Journey time
    - Preferred destinations (which are the ports of greatest interest)

- Respondent profile

To carry out this work a series of stages has been followed. This will be defined and summarized below:

2.1 (a) Obtain preliminary information: collect data on the resident population, as well as the tourist population received.

2.1 (b) Definition of objectives under the consensus of all the researchers participating in the project: definition of the variables from which information is sought, as well as determination of the universe of the study, the type of sampling to be used, and the optimal sample size.

2.1 (c) Preparation of the questionnaire for the analysis of the resident population. For the tourist population, given the impossibility of being able to access this public by the dates of execution, a script was prepared to carry out an exploratory investigation with a panel of experts.

2.1 (d) Fieldwork: to carry out the surveys an external company was hired; the same questionnaire was also used to obtain answers online, and in relation to the panel of professionals in the sector the interviews were conducted by the researchers themselves.

2.1 (e) Tabulation and registration of the data: design of a database to record the information from the surveys previously prepared.

2.1 (f) Analysis and interpretation of the results: this work was carried out with the support of statistical techniques.

2.1 (g) Preparation of the final report.

2.2 DESIGN OF TRANSPORT LINES

The objective of this project activity is to establish maritime routes between the regions of Cádiz, Huelva, and Algarve, for which it is necessary to study different factors that will determine which routes are the most feasible according to the meteorological limitations of the area and existing infrastructure in the different ports of the study regions and their peculiarities.

The result of these activity is designed to determine the different possible lines (using different ports for of the three areas under study), and the capacity, service speed, and number of boats required to comply with the demand study requirements. All this will help select the best connections in terms of economic and environmental viability.

2.3 DESIGN OF SUSTAINABLE VESSELS

This activity has begun with the development of a parametric statistical study with the main characteristics of similar ships. This study has been used to analyse the relationships between the passenger capacity and speed versus the installed propulsion power. With this information, it has been possible to make a preliminary dimensioning of the ships to be used in the analysed potential maritime lines. This dimensioning, in terms of length and propulsion power, has been used to carry out a study of consumption and emissions of all proposed lines, so that both economic and environmental viability could be determined and thus select the most optimal design alternative to then deepen and define the boat.

Note that the design of these new ferries must face several challenges, among which are competing in time with the existing land transport in the area as well as being efficient and ecological, providing comfort and safety to passengers.

2.4 ENVIRONMENTAL STUDY

The methodology used at this stage has started with a general study of the polluting factors associated with maritime traffic. Similarly, environmental regulations affecting international maritime navigation have also been considered. We will then proceed to establish the most important potential environmental impacts linked to such mode of transport.

To conclude, all this information and regulations have been applied to the transport system of the project, in addition to making an environmental comparison of the carbon footprint for the three means of transport considered in the study (car, bus, and ship).

To compare the efficiency of the two different transport systems under discussion (land and sea transport), and bearing in mind that any development of existing propulsion power systems such as the use of electric engines, the use of natural gas or hydrogen as fuel or other future development, could be installed in vehicles used for both maritime and inland transport, it was decided to make the comparative analysis using similar fuel and propulsion engines in all vehicles. In our case, similar diesel engines were theoretically "installed" in some standard bus, car, and ferry, and the transport efficiency comparison was then made in terms of installed power consumed per passenger, and therefore, fuel and $CO2$ emissions made per passenger. This comparison has been based on the hypothesis that there is a direct bus connection between the study populations, as well as that the occupancy of the different transports are 100%, and that the propulsion system used by them is the same.

2.5 ECONOMIC ANALYSIS

In this activity, an estimation of the costs that would involve the implementation of the maritime lines of study has been carried out. For this purpose, the costs associated with the different phases of the exploitation of this new transport system have been established, such as the acquisition and/or rental of the vessels, maintenance, personnel on board and land, cleaning, etc.

Taking into account the maximum price to be paid for the sea route established in the demand study previously carried out and the estimated costs according to the information obtained, a cost-benefit analysis of the different maritime lines considered in the investigation has been carried out.

All these data, together with more information, will be used to study the economic impact that the implementation of this project could have on the area of action.

2.6 OPTIMIZATION

In this activity, the optimization and improvement of the transport system is also carried out based on the results obtained and the final conclusions of the entire study being carried out.

Note that in addition to the main activities mentioned here, other phases have been also carried out and are being carried out in the project, such as the preparation, management, coordination, of the project, and communication about the project.

3. RESULTS

3.1 DEMAND STUDY AND ANALYSIS

After having carried out the surveys and having collected all the information in relation to the demand that would exist in the area, it was determined that leisure would be the main reason for the passenger displacements, both during vacation periods and, to a lesser extent, throughout the year. In other words, maritime transport in not seen by the potential users as a regular transport system but as an attractive new mode of transport used to connect touristic regions or populations. There are preferences for traveling with family or group and, if possible, in the morning. With respect to travel conditions, the price variable is the most important, ranging from € 20 - € 40 for a single ticket each way.

Furthermore, the maximum travel time allowed by the respondents is three hours for the Cádiz-Huelva connection and four hours for Cádiz-Algarve or Algarve-Huelva, allowing an increase of one hour on all trips provided they are accompanied by playful activities. The time taken to travel from the origin to the departure port and from the arrival port to the destination point has also been taken into account; this should be between thirty and forty-five minutes.

The ports preferred by respondents in the province of Cádiz are Cádiz capital, Puerto de Santa María, and Sanlúcar; while for the province of Huelva the preferred ports are Huelva capital, Matalascañas, Punta Umbría, and Isla Cristina, and in the Algarve area the ports of Faro, Lagos, and Albufeira stand out.

Table 1 summarizes the characteristics that, according to the demand study, the maritime connections between the different areas should have.

Provinces	Passengers/day	Trip duration (h)	Maximum ticket price (€)
Cádiz – Huelva	1,614	3	30
Cádiz – Algarve	1,512	4.5	40
Algarve – Huelva	886	4.5	30

Table 1. Summarised result of demand study

Note that the season of the year and the number of passengers per day with which the study has been carried out are those considered during the holiday period, that is, from June to September since, according to the demand analysis, it is when greater demand would be.

3.2 DESIGN OF TRANSPORT LINES

Once the study of the different nautical factors that determine the viability of the possible maritime lines had been carried out, it has been proven that navigation in the Gulf of Cádiz is fully feasible. According to the climatological study, it is expected that on some occasions the proposed connections would be affected by wind and waves, having to suspend their operation, although it should be noted that in the summer this possibility would be very low. On the other hand, there are ports that are conditioned by the limitations of navigation speed in inland waters, due to their physiognomy, compatibility with other port activities, or environmental protection as *Faro in* Algarve or *Muelle de Levante* in Huelva. In addition, we must also take into account that there are ports in the area of action without the necessary infrastructure to operate passenger ships. For these and other reasons, some ports have been excluded in the study, thus discounting the maritime lines in which they intervened.

In the same way, other ports of interest have been taken into account, both those detected in the demand study that were not previously contemplated, as well as those considered by the technically viable researchers from the nautical point of view, such as Vilamoura and Portimao in Portugal, Mazagón in Huelva, and Chipiona and Rota in Cádiz, being defined as follows. In Table 2 we can see the maritime lines used in this study.

Provinces	Lines	Populations		Boat capacity	Service Speed (knots)	No. of journeys
Cádiz-Huelva	Line 1-a	Cádiz	Huelva	400	28.8	2
	Line 1-b	Cádiz	Huelva	200	28.8	4
	Line 2	Cádiz	Mazagón	200	24.0	4
	Line 3	Rota	Mazagón	200	26.1	4
	Line 4-a	Chipiona	Mazagón	200	22.2	4
	Line 4-b	Chipiona	Mazagón	100	22.2	8
Cádiz-Algarve	Line 5-a	Cádiz	Portimao	750	30.3	1
	Line 5-b	Cádiz	Portimao	375	30.3	2
	Line 6-a	Cádiz	Vilamoura	750	24.0	1
	Line 6-b	Cádiz	Vilamoura	375	24.0	2
Algarve-Huelva	Line 7	Portimao	Huelva	220	27.6	2
	Line 8	Portimao	Mazagón	220	24.9	2
	Line 9	Vilamoura	Huelva	220	29.7	2
	Line 10	Vilamoura	Mazagón	220	26.2	2

Table 2. Maritime lines of study

Since the demand study only contemplated the use of the ports of Cádiz, Huelva, and Faro, multimodal connections have been established that include the transfer in shuttle buses from the alternative ports (Chipiona, Mazagón, and Rota) to the main ports of Cádiz and Huelva and vice versa, in order to be able to use the demand data and connection times required in the analysis of these alternative lines.

Also included in this nautical research is the study of the applicable regulations regarding maritime safety and pollution prevention that the vessel has to comply with have also been considered.

3.3 DESIGN OF SUSTAINABLE VESSELS

As we have said before, in this activity an initial parametric statistical study has been carried out, which has helped in the preliminary design of the vessels that could be used in the study transport system. To analyse the most optimum condition of the environmental study, the best existing ferries, in terms of lower installed propulsion power per passenger, were selected according to the capacity and speed required by the demand study. Table 3 shows the consumption per passenger and emissions of each of the analysed lines in case these existing ferries were used.

Provinces	Lines	Populations		Power (kW)	Total diesel consumption (litres)	CO_2 emissions (kg)/pax (g)
Cádiz-Huelva	Line 1-a	Cádiz	Huelva	1,940	1,013.4	6.6
	Line 1-b	Cádiz	Huelva	1,940	1,106.8	14.4
	Line 2	Cádiz	Mazagón	1,340	617.2	8.8
	Line 3	Rota	Mazagón	1,340	538.2	8.6
	Line 4-a	Chipiona	Mazagón	1,340	465.1	6.7
	Line 4-b	Chipiona	Mazagón	684	251.6	8.1
Cádiz-Algarve	Line 5-a	Cádiz	Portimao	5,386	4,130.9	17.8
	Line 5-b	Cádiz	Portimao	2,459	2,249.1	17.6
	Line 6-a,	Cádiz	Vilamoura	5,386	3,415.3	14.7
	Line 6-b	Cádiz	Vilamoura	2,459	1,687.4	13.2

Algarve-Huelva	Line 7	Portimao	Huelva	1,574	1,370.7	15.8
	Line 8	Portimao	Mazagón	1,574	1,230.0	14.6
	Line 9	Vilamoura	Huelva	1,574	1,101.5	13.1
	Line 10	Vilamoura	Mazagón	1,557	1,014.2	12.6

Table 3. Consumption and emissions of existing vessels

As we can see in Table 4, the maritime line with lower consumption per passenger compared with bus and therefore, with greater environmental viability, is the one that connects the provinces of Cádiz and Huelva. For this reason, the optimized vessel model for the exploitation of regular passenger transport of this line is being designed. Meanwhile, for the other maritime connections of the project (Cádiz-Algarve and Huelva-Algarve), the development of recreational/tourist vessels in this geographical area is being designed in collaboration with the Coventry University.

It should be noted that the design of the vessel for the Cádiz - Huelva connection has to be a hydrodynamically and structurally optimized design to be able to achieve a more efficient maritime transport than a hypothetical direct public transport line that is now non-existent, as previously mentioned.

3.4 ENVIRONMENTAL STUDY

As we have seen in the previous sections, the study of fourteen maritime lines has been carried out. This study includes a comparison, which we can see in table 4, of fuel consumption and CO_2 emissions between the maritime transport system of this project and the existing land transport in the area, resulting in emissions from passenger, compared to the private, vehicles are smaller for maritime connections between Cádiz and Huelva, while these are increased for the connection of Cádiz and the Algarve and of Huelva with the Algarve. If we make the comparison with an ideal connection in public transport, that is, if there is a direct bus line between the main connection ports, all the maritime lines studied would have a higher consumption and emissions per passenger.

Provinces	Lines	Populations		Maritime Line vs Bus	Maritime Line vs Car
				CO_2 Emissions (%)	CO_2 Emissions (%)
Cádiz-Huelva	Line 1-a	Cádiz	Huelva	177	70
	Line 1-b	Cádiz	Huelva	385	153
	Line 2	Cádiz	Mazagón	234	93
	Line 3	Rota	Mazagón	229	91
	Line 4-a	Chipiona	Mazagón	180	72
	Line 4-b	Chipiona	Mazagón	216	86
Cádiz-Algarve	Line 5-a	Cádiz	Portimao	474	189
	Line 5-b	Cádiz	Portimao	470	187
	Line 6-a	Cádiz	Vilamoura	392	156
	Line 6-b	Cádiz	Vilamoura	352	140
Algarve-Huelva	Line 7	Portimao	Huelva	420	167
	Line 8	Portimao	Mazagón	389	155
	Line 9	Vilamoura	Huelva	348	139
	Line 10	Vilamoura	Mazagón	335	134

Table 4. Environmental comparison in terms of CO_2 emissions

Within all lines analysed, the one with the highest energy efficiency and, therefore, the greatest environmental viability is the one that connects the provinces of Cádiz and Huelva through the ports of Chipiona and Mazagón. This would enable the implementation of regular passenger transport. While, for the other maritime lines (Cádiz - Algarve and Huelva - Algarve), the implementation of a recreational/tourist transport system is being considered.

3.5 ECONOMIC ANALYSIS AND OPTIMIZATION

From the preliminary cost-benefit analysis that has been carried out, there are several connections that are more economically viable, among which are some lines that link Cádiz-Huelva and Cádiz-Algarve. This study is being analysed in order to improve the results. In addition to this an analysis will be made of the economic impact that will occur in the area of action in the event that this new maritime passenger transport system is implemented.

Once we have determined all the factors that allows us to compare the technical, economic, and environmental viability of the different maritime connections studied, we will be able to select the optimum and best option to be publicly recommended.

4. CONCLUSIONS

In the demand study we can observe that there is a great interest for local citizens in possibly using this new mean of transport, mainly in a playful/tourist way during holiday periods. So much so that the business association of Chipiona (ACITUR) has created a civic platform (Uniendo Provincias) in which the implementation of this maritime connection between the provinces of Cádiz and Huelva is promoted. This interest is in addition to other groups, such as the "Mesa de la Ria" in Huelva, which also have been promoting this as a political and civil demand for a number of years.

From the study of the transport lines, we can deduce that the navigation in the geographic area of the research, the Gulf of Cádiz, is totally feasible and that the lines with greater probability of execution have been selected for later analysis.

In the design of sustainable vessels work is being done on the optimization of the vessel that would connect the provinces of Cádiz and Huelva, through the ports of Chipiona and Mazagón. This line is, according to the environmental impact study, carried out, the one which presents greater energy efficiency and greater environmental viability for regular passenger transport.

For the design of all the vessels of the project, including those for recreational purposes, the powers and, therefore, the consumption and emissions, determined by the design of the lines and by the parametric statistical study initially carried out, are being taken into account.

The results, for now, of the study being carried out, indicate that the most efficient line is 4, Chipiona - Mazagón. So the design of an optimized and sustainable vessel is focusing on the Cádiz-Huelva connection for regular passenger transport through these ports, together with the analysis of cost-benefits and economic impact in the area of the Gulf of Cádiz.

5. ACKNOWLEDGEMENTS

The authors would like to thank the European Regional Development Fund (ERDF) of the European Union (EU) for co-financing the project "ESPOmar" under the "INTERREG POCTEP 2014-2020 program". We would also like to thank to all researchers and staff of the different public organizations involved on the project for their dedication and professionalism:

University of Cádiz: Antonio Arcas de los Reyes, Antonio Querol Sahagún, Antonio Rafael Peña Sánchez, Begoña Flethes Bernal, Diego Sales Lérida, Diego Sales Márquez, Emilio Bravo López, Emilio Rodríguez Díaz, Enrique Nebot Sanz, Francisco Piniella Corbacho, Germán Jiménez Ferrer, Jose Angel Llamas Alfaro, José Ruiz Chico, Juan Antonio Palacios García, Juan Carlos Rasero Balón, Juan Ignacio Alcaide Jiménez, Juan Moreno Gutiérrez, Luis Alberto Fernández Güelfo, María del Carmen Pérez Peña, María José Legáz Almansa, Mercedes Jiménez García, Nieves Endrina Sánchez, Patricia Sánchez Garrido, Rafael Jiménez Castañeda, Raquel Núñez- Barranco González-Elipe, Ruth García Llave, Sebastián Sotomayor González, Vanesa Durán Grados.

University of Huelva: Alfonso Vargas Sánchez, José Albendín Moya.

University of Algarve: Cláudia Almeida, Fernando Perna, Francisco Serra, Margarida Santos, Maria João Custódio.

Andalusian Public Port Agency: Diego Anguís Climent, Fernando Copado.

6. REFERENCES

ADOLF K.Y., NG JASON, MONIOS CHANGMIN JIANG "Maritime Transport and Regional Sustainability", *Elsevier*, 2019

GUNES U., BEJAN A., SAHIN B., "The evolution of air and maritime transport", *Applied Physics Reviews 6*, 2019

CHRISTODOULOU A., WOXENIUS J., "Sustainable Short Sea Shipping", *Sustainability*, 2019

NTANOS S., PARAVANTIS J., SAMBRACOS E., "Energy Consumption and Carbon Dioxide Emissions of a Suburban Coastal Transport System", *Munich Personal RePEc Archive*, 2008

PARAVANTIS J.A., GEORGAKELLOS D., "Trends in Energy Consumption and Carbon Dioxide Emissions of Passenger Cars and Buses", *Technological Forecasting and Social Change, 74, 682-707*, 2007

SAMBRACOS E., "Investment Evaluation of a Suburban Coastal Transport System", *10th International Conference of IFORS (International Federation of Operations Research Societies), 1-3 April 2001, Athens, Greece*, 2001

COTTAM H., ROE M., CHALLACOMBE J., "The impact of transitional changes of maritime passenger transport on Croatian tourism development", *Maritime Policy & Management 34(3):185-202*, 2007

PINIELLA F., RASERO JC, QUEROL A., "Maritime Passenger Transport as an Urban and Interurban Alternative on the River Guadalquivir: GuadaMAR", *Maritime Transport VIAt: BarcelonaVolume: 1*, 2014

7. AUTHORS' BIOGRAPHY

F. Piniella is Rector of the University of Cádiz, and Full Professor in the Department of Sciences and Techniques of Navigation and Naval Constructions at the University of Cádiz, Spain.

A. Querol is Principal Researcher of ESPOmar Project and Associate Professor and Director of the Departmental Section of the Department of Sciences and Techniques of Navigation and Naval Constructions in the University of Cádiz, Spain.

P. Sánchez is ESPOmar Researcher in the Department of Sciences and Techniques of Navigation and Naval Constructions at the University of Cádiz, Spain.

J.C. Rasero is Associate Professor in the Department of Sciences and Techniques of Navigation and Naval Constructions, and Director of the Central Service for Resources and Nautical Infrastructures at the University of Cádiz, Spain.

L.A. Fernández is Associate Professor in the Department of Environmental Technologies, and Deputy Director of the Verinsur Chair in Environmental Research and Technology at the University of Cádiz, Spain.

S. Sotomayor is Associate Professor in the Department Financial Economics and Accounting, Director of Departmental Section of the Department Financial Economics and Accounting in Functions, and Vice-Dean of Academic Management and Institutional Relations of the Faculty Social Sciences and Communication at the University of Cádiz, Spain.

D. Sales is Director of the Department of Environmental Technologies at the University of Cádiz, Spain.

A. Vargas is Full Professor in Business Organization at the University of Huelva, Spain.

F. Perna is Coordinating Professor of Economics and Tourism at the University of Algarve, Portugal.

M.C. Pérez is Interim Lecturer in the Department of General Economy at the University of Cádiz, Spain.

DESIGN-DRIVEN INNOVATION: INNOVATIVE AND CULTURALLY SPECIFIC RETRO-FUTURISTIC DESIGN APPROACH TO INTERIOR DESIGN AND BESPOKE FURNITURE FOR A FLAMENCO VENUE WITHIN A 280M SUSTAINABLE LUXURY TRANSATLANTIC SUPERLINER CONCEPT "LA TRASATLÁNTICA"

S. McCartan, EBDIG, FAH, Coventry University, UK
D. Moreno Nieto, EBDIG, Cádiz University, Spain

SUMMARY

This paper reports on the development of an innovative and culturally specific retro-futuristic design approach to interior design and bespoke furniture of a Flamenco venue on a transatlantic superliner. It is developed within the context of a Design-Driven Innovation concept proposal of a 280m pentamaran high speed transatlantic superliner, inspired by the design meaning of the Normandie for the region of Andalucía, as a homage to the cultural heritage of Compañía Transatlántica Española (CTE) or "La Trasatlántica". The interior design and furniture design processes were stimulated by the innovative manufacturing capabilities of 3D printing technology. Luxury has undergone a paradigm shift from materialism to experiential value. Millennials crave new and unique curated experiences to see a place that few or no people have previously seen. Sustainable luxury is the new luxury agenda; it offers the opportunity of a new design meaning for the next generation of superliners through a holistic approach encompassing a range of issues from addressing pollution to societal contributions. The use of bioplastics in 3D printing and natural fibre-based materials engages bespoke furniture design with the sustainable luxury agenda.

1. INTRODUCTION

Historically, due to its innovative technical design and luxurious interiors the Normandie is considered the greatest of all superliners. The concept of Normandie was developed in the Roaring Twenties as a next generation vessel for upper-class Americans tourists who travelled to Europe for alcohol-fuelled fun during prohibition. She was designed to represent France in the nation-state contest of the great liners and was built in a French shipyard using French parts. The luxurious interiors were designed in Art Deco and Streamline Moderne style, with most of the public space assigned to first-class passengers. This paper reports on the development of an innovative and culturally specific retro-futuristic design approach to the interior design including bespoke furniture design of a Flamenco venue on a transatlantic superliner. Developed within the context of a Design-Driven Innovation concept proposal of a 280m pentamaran high speed transatlantic superliner, inspired by the design meaning of the Normandie for the region of Andalucía, as a homage to the cultural heritage of Compañía Transatlántica Española (CTE) or "La Trasatlántica". The interior and furniture design processes used were stimulated by the innovative manufacturing capabilities of 3D printing technology. Luxury has undergone a paradigm shift from materialism to experiential value. Millennials crave new and unique curated experiences, to see a place that few or no people have previously seen. Sustainable luxury is the new luxury agenda; it offers the opportunity of a new design meaning for the next generation of superliner through a holistic approach encompassing a range of issues from addressing pollution to societal contributions. The use of bioplastics in 3D printing and natural fibre materials engages bespoke furniture design with the sustainable luxury agenda.

Fig. 1: EBDIG 280m Pentamaran High Speed Transatlantic Superliner Concept

This project builds upon the work of McCartan, Crotty, and Verheijden [1], on the 280m pentamaran high speed transatlantic superliner proposal, shown in Figure 1. Their concept engaged in Design-Driven Innovation to develop a new market sector for high speed multifunctional vessels to compete with both air freight and business class air travel, in addition to performing the role of a superliner cruise ship. This design proposal offered the business traveller a personalised office space with global connectivity to make the journey a seamless extension of the working environment. This was specifically targeted at the creative design industries, where staff are used to large design office spaces for collaborative working. There were also conference facilities and exhibition spaces to facilitate product launches and client negotiations. The logistics role of the vessel, through competing with air freight and being assigned all of the CO2 footprint, gave the cruising passengers and business travellers a zero-carbon footprint. The vessel was based on the BMT ADX Express fast container ship Lo-Lo platform, capable of a speed of 37

Knots. This offers a significant reduction in fuel consumption compared with conventional mono-hull platforms when traveling at high speed.

1.1 Design-Driven Innovation

The process of Design-Driven Innovation is an exploratory research project which aims to create an entirely new market sector for a given product through changing the design meaning the user has for the product. It occurs before product development. It is not the fast and creative brainstorming sessions that are typical of concept generation but rather a design investigation, similar to technological research. It is the development of a design scenario through engaging with a range of interpreters in both technology and cultural production. Knowledge is generated by immersion in the design discourse of the interpreters' groups. The interaction between innovation of design meaning and technology innovation can transform the market within an industry and even create new market sectors. The successful interaction between design-driven and technology-push innovation is called a technology epiphany. It creates a market leader and, in some cases, a completely new market sector. It is the basis for successful products such as the Apple iPod. [2]

Feadship engage in the DDI discourse through Future Concept Feadships as unique projects taken from the concept through the engineering stages and presented with a model and documentation at the Monaco Boat Show. This enables them to engage in DDI with stakeholders, especially important as a new concept for an owner will take four years before it is finished and presented to the market as a completed yacht. Thus, the Future Concept Feadships produces a strong marketing narrative to enhance client meetings. The Shipyard 4.0 is a new concept of the connected Industry 4.0 which emphasizes leveraging the potential of new technologies and design practices. It seeks to combine international best practices within those areas, both in the field of shipbuilding and in other industries, to create a state-of-the-art innovation approach such as DDI. The significant opportunity for bespoke furniture design using 3D printing, and the lighting design opportunities of the clear and translucent printing materials, form the basis of this project. This project was collaboration between Cadiz University Department of Naval Architecture, Cadiz University Department of Industrial Design, and the Automotive & Transport Design Course at Coventry University.

1.2 History of Trasatlántica

La Trasatlántica was founded in Cuba by Antonio Lopez y Lopez in 1850. In 1861 López won the concession of passenger and mail transport contracts between Spain, Puerto Rico, Cuba, and Santo Domingo. To this concession was added the transfer of soldiers to Santo Domingo (1863-1865) and to Cuba during the War between 1868 and 1878. The size of the fleet grew with the company's expansion. In 1878 the shipping company had thirteen steamers for its regular lines from Santander and Cádiz to Cuba and Puerto Rico. That same year King Don Alfonso XII granted Antonio López y López the title of Marques de Comillas. The Transatlántica Company was established in Barcelona in 1881. It started with a service between Alicante, Barcelona, and Marseille, and developed a postal service between Spain, Cuba, Puerto Rico, and Santo Domingo in 1861. In 1881 the parent company was transferred from Cuba to Barcelona, with the transatlantic terminal being moved from Cádiz to Barcelona. Completely new lines were progressively developed, including a mail service between Liverpool, Spain, and the Philippine Islands via the Suez Canal. Another line wnet from Cádiz to La Guaira, Puerto Cabello, Savanilla, Cartagena, and Panama, and there was one from Barcelona to Fernando Po. At the end of the century a relay service between Havana and New York, which had started in the 1880s, had evolved into a direct route from Spain to New York, Havana, and Vera Cruz.[3]

Between 1881and 1883 the following new vessels were purchased: Antonio López, Cataluña, and The City of Santander. The City of Santander made several annual trips to South America from 1889 until being grounded in 1895. Having made over ten round trips to South America the Antonio López was one of the five ships that the Transatlantic Company lost during the Spanish-American war. The Antonio López sailed from Buenos Aires to Barcelona on 2 January 1898, making it her last or penultimate trip ot South America. A twin, Cataluña, had made a trip to Buenos Aires in 1892. During the war, eluding the blockade imposed by the American navy, she made several successful trips, mainly to Cuba, carrying provisions, armaments, and troops. On 16 June 1898 she had left the port of Cádiz with a large cargo of weapons and other supplies for San Juan. She was attacked on 28 June by the Yosemite as she approached the port of San Juan. The cargo was recovered and San Juan was reinforced resulting in the American forces having to abandon their plans to attack the city directly. The last vessel known to have made a trip to South America before the war was the City of Cadiz. Between 1893 and 1894 she made four round trips to Buenis Aires. [3]

The Cuban Revolution in 1896 resulted in the transport of more than 200,000 Spanish troops to the island. All available Spanish passenger ships were chartered by the Government of Spain. After the war, Fr. De Satrústegui made a number of trips between Liverpool, Barcelona, the Suez Canal, and Manila, just like another former British Indian ship, Taroba, which was renamed Leon XIII. Part of her duties at this time was the evacuation of Spanish troops from the Philippine Islands. In the course of time Fr. De Satrústegui left Liverpool for Barcelona. On 24 September 1899 she set sail for Genoa but when approximately forty miles off the coast of Valencia she collided with the Bassano of the Wilson Line, causing great damage to her bow. Having only suffered minor damage the Spanish ship returned to Barcelona after stopping in Genoa and departed again on 27 November to Buenos Aires. Fr de Satrústegui made sporadic trips to South America between 1900 and

1901, made four or five trips in both 1902 and 1903, and from 1904 to 1913 she maintained a regular monthly service together with Leon XIII, who made at least four trips in 1901. [3]

A short-lived service was introduced by the Mexico, formerly the British Trentham Hall, in 1875, which left Liverpool on 23 March 1900 headed for the Strait of Magellan and Valparaíso, on the outward journey sailing via Montevideo and on the return trip via Rio de Janeiro. It was followed by the San Agustin, which left Liverpool on 24 May 1900, returning there on 15 September with an itinerary very similar to its predecessor except that it stopped in Rio de Janeiro on the trip one way and in Montevideo on the return trip. The second trip of the Mexico began in Liverpool on 22 July and that of San Agustin on 24 September.

During her return trip the San Agustin was wrecked on the Strait of Magellan but there were no casualties. On her third trip from Liverpool the Mexico did not continue beyond Buenos Aires. When sailing to Vigo on the return trip she was stranded off the coast of Portugal. She was a total write-off but there was no loss of life. [3]

By 1912 more than twenty years had passed since the company had commissioned a new transatlantic vessel. Therefore it was a great event when, on 26 September 1912, Swan, Hunter & Wigham Richardson of Wallsend-on-Tyne launched the Queen Victoria Eugenia a vessel with four propellers, especially so as two days later Wm. Denny & Bros. launched the Infanta Isabel de Borbon, a vessel with three propellers. Both had been specially built for the route to the Río de la Plata. Both had a service speed of 17 knots with a capacity for 250 first, 100 second, and 75 third class passengers. The inaugural trips began from Barcelona on 12 March and 4 April 1913 respectively. Previously they had stopped in Genoa, then in Malaga, Cádiz, Tenerife, and Montevideo en route to Buenos Aires. Within a few months Fr de Satrústegui and Leon XIII left the Mediterranean and sailed to the Río de la Plata from Bilbao, Santander, and Coruña, and often stopped at additional ports. The First World War affected the departure schedules of the Transatlantic Company between August 1914 and November 1918 less than any other major line: Queen Victoria Eugenia and Infanta Isabel de Borbon made a number of trips between Spanish and New York ports during the war. [3]

For most of the war the Fr. De Satrústegui and Leon XIII continued their travels from Bilbao, Santander, and Coruña to Montevideo and Buenos Aires, while the Buenos Aires made one or two trips to South America, including one from New York to Buenos Aires. A two-propeller vessel whose career began in 1892 as the HH Meier of Norddeutscher Lloyd and continued in 1901 as the Manuel Calvo of the Transatlantic Company, sailing from Barcelona to Buenos Aires on 10 October 1917. After the war the activities of the Queen Victoria Eugenia and the Infanta Isabel de Borbon continued successfully but with a more extended itinerary for part of the time: Barcelona, Almería, Málaga, Cádiz, Tenerife, Rio de Janeiro, Montevideo, and Buenos Aires. As a result of the opening of the Panama Canal a new service was introduced from Barcelona to Puerto Rico, La Guaira, Puerto Colombia, Panama Canal, and Valparaíso. One of the vessels used was Leon XIII. [3]

Fig. 2: Art Deco CTE poster [3]

In 1931 the new republican Government of Spain was recognized by the US. One consequence of which was the renaming of ships. Alfonso XIII (III) was renamed Havana, while Queen Victoria Eugenia and Infanta Isabel de Borbon were renamed Argentina and Uruguay respectively. Both sunk during the Spanish Civil War of 1936 to 1939. On 8 May 1932 it was announced in Barcelona that the service to the Rio de la Plata would be suspended given the inability of the Spanish Government to maintain its subsidy. The Uruguay had already made her penultimate departure from Buenos Aires to Barcelona on 30 April 1932. There was a large turnout of Spaniards in the port of Buenos Aires to see off the Argentina when she set sail for Spain on 2 June. From then on it was reported that ships would travel to Cuba, Central America, Venezuela, and Colombia. In 1939, after the Civil War, the Company's parent company was transferred from Barcelona to Madrid. Several new passenger ships were built after World War II, and were widely used, but not to the Río de la Plata. Passenger ships began to be supplanted by airliners in the late 1960s, and the company ceased trading in December 1974. [3]

2. SPANISH REGIONAL CULTURE AND DESIGN

The Generation of 1898 was a group of novelists, poets, essayists, and philosophers active in Spain at the time of the Spanish–American War (1898), who were committed to cultural and aesthetic renewal, and were associated with modernism. The name Generation of 1898 was coined by José Martínez Ruiz, in his 1913 essay entitled 'La generación de 1898', alluding to the moral, political, and social crisis in Spain produced by the loss of the colonies of Cuba, Puerto Rico, the Philippines, and Guam after Spain's defeat in the Spanish–American War. The intellectuals included in this group are known for their criticism of the Spanish literary and educational establishments. They saw these being characterised by conformism, ignorance, and a lack of any true spirit. Their criticism was coupled with and heavily connected to the group's dislike for the Restoration Movement adopted by the Spanish government. [4]

El Concurso del Cante Jondo (Contest of the Deep Song) was a fiesta of flamenco arts, music, song, and dance held in Granada in 1922. Conceived and initiated by composer Manuel de Falla it was strongly support from the poet Federico García Lorca. The two-day evening event was held outdoors at the Alhambra. The show included the best of well-known flamenco artists. The aim of the Concurso was not merely to celebrate the music, but to culturally elevate it. Enlisting the support of Spanish intellectuals was considered crucial to counter the then-current flood of antiflamenquismo feeling of the Generation of 1898. These Spanish reformers were activists in the widespread effort to modernize and transform Spain and its culture, yet in the process they judged flamenco harshly as being frivolous and regressive. For the Concurso to succeed Falla aimed not only to produce a stellar musical event but also to attract an influential audience as witnesses. The concurso obviously attracted flamenco musicians and aficionados but Falla also wanted to impress the luminaries, influential figures in the wider worlds of art and culture, to ensure that after the event they would sing the praises of cante jondo. The Spanish press generally spoke in praise of the Concurso. A Madrid magazine published soon after the event described the Concurso as 'unforgettable', with its alternating displays of yearning, vehemence, superstition, or fervour, 'a simple seduction of sound, rhythms linear in the flesh'. Following this there was a steady rise in the status of flamenco among the cultural and intellectual leaders of Spain.[5]

2.1 Catalan Modernism and Gaudi

Catalan Modernism was an art and literature movement associated with the search of a new entitlement of Catalan culture. Its main form of expression was in architecture. Although Modernisme was part of a general trend that emerged in Europe at the turn of the twentieth century, in Catalonia the trend acquired its own unique personality. Modernisme's distinct name comes from its special relationship, primarily with Catalonia and Barcelona, which were intensifying their local characteristics for socio-ideological reasons after the revival of Catalan culture and in the context of spectacular urban and industrial development. It is equivalent to the following fin de siècle art movements: Art Nouveau in France and Belgium; Jugendstil in Germany; Vienna Secession in Austria-Hungary; Liberty style in Italy; and Modern Style in Scotland. It was a search for a particular style for Catalonia drawing on medieval and Arab influences. As well as combining a rich variety of historically derived elements, it is characterized by the predominance of the curve over the straight line, by rich decoration and detail, by the frequent use of organic motifs, the tendency for asymmetry, and a refined aestheticism and dynamic shapes. While Barcelona was the centre of Modernista construction, the Catalan industrial bourgeoisie built industrial buildings and summer residences. [6]

Antoni Gaudí was a Catalan architect known as the greatest exponent of Catalan Modernism. His works have a highly individualized style. Most are to be found in Barcelona, including his main work, the church of the Sagrada Família. Gaudí's work was influenced by his passions in life: architecture, nature, and religion. He considered every detail of his creations and integrated into his architecture such crafts as ceramics, stained glass, wrought ironwork forging, and carpentry. He also introduced new techniques in the treatment of materials, such as trencadís which used waste ceramic pieces. Under the influence of neo-Gothic art and Oriental techniques Gaudí became part of the Modernista movement which was reaching its peak in the late nineteenth and early twentieth centuries. His work transcended mainstream Modernisme culminating in an organic style inspired by natural forms. In 1904 Josep Batlló hired Gaudí to design his home. The building was completed and refurbished in 1906. He completely changed the main apartment which became the residence for the Batlló family. He expanded the central well in order to supply light to the whole building and also added new floors. [6]

The local name for the building is Casa dels ossos (House of Bones) as it has a visceral, skeletal, organic quality. The ground floor, in particular, is rather astonishing with tracery, irregular oval windows, and flowing sculpted stonework, as shown in Figure 3. It seems that the goal of the designer was to avoid straight lines completely. Much of the façade is decorated with a mosaic made of broken ceramic tiles (trencadís) that starts in shades of golden orange moving into greenish blues. The roof is arched and was likened to the back of a dragon. A common theory about the building is that the rounded feature to the left of centre, terminating at the top in a turret and cross, represents the lance of Saint George (patron saint of Catalonia), which has been plunged into the back of the dragon. The noble floor is larger than seven-hundred square meters. It is the main floor of the building. The noble floor is accessed through a private entrance hall that utilizes skylights resembling tortoise shells and features vaulted walls in curving shapes. On the noble floor there is a spacious landing with direct views to the blue tiling of the building well. On the Passeig de Gracia side is Mr. Batlló's study, a dining room, and a secluded spot for courting couples, decorated with a mushroom-shaped fireplace. The elaborate and animal-like décor continues throughout the whole noble floor. As shown in Fig.4, the building's façade has three distinct sections which are harmoniously integrated. The lower ground floor with the main floor and two first-floor galleries are contained in a structure of Montjuïc sandstone with undulating lines. The central part, which reaches the last floor, is a multicoloured section with protruding balconies. The top of the building is a crown, like a huge gable. The top displays a trim of ceramic pieces. [6]

Fig. 3: Casa Batlló interior view of the ground floor

Sagrada Família is a large unfinished Roman Catholic minor basilica in Barcelona. On 19 March 1882 construction of the Sagrada Família began under the direction of architect Francisco de Paula del Villar. When Villar resigned in 1883, Gaudí took over as chief architect, transforming the project with his architectural and engineering style, combining Gothic and curvilinear Art Nouveau forms. Gaudí devoted the remainder of his life to the project, and he is buried in the crypt. At the time of his death in 1926 less than a quarter of the project was complete. The columns of the interior, shown in Fig.5, are a unique Gaudí design. Besides branching to support their load, their ever-changing surfaces are the result of the intersection of various geometric forms. The simplest example is that of a square base evolving into an octagon as the column rises, which then becomes a sixteen-sided form, and eventually leads to a circle. This effect is the result of a three-dimensional intersection of helicoidal columns (for example a square cross-section column twisting clockwise and a similar one twisting counter-clockwise). Essentially none of the interior surfaces are flat; the ornamentation is comprehensive and rich, consisting in large part of abstract shapes which combine smooth curves and jagged points. Even detail-level work such as the iron railings for balconies and stairways are full of curvaceous elaboration. [6]

The sustainability of Gaudí's architecture in some of his most significant works, such as the Batlló House and Sagrada Familia, is revolutionary as it demonstrates bioclimatic architecture at the beginning of the twentieth century when the concept of 'bioclimatic design' did not exist yet and introduces the theories of biomimetic architecture of the twenty-first century. Accordingly, it can be considered that Gaudí was, really, a precursor of sustainability and biomimetics in architecture, almost a century in advance of the formalization of these design practices. The architect Juan Bergós Massó, a student of Gaudi, commented that Gaudi showed that artistic inspiration was always the result of intense, methodical, and disciplined work. Where the organic forms, so unusual and surprising, are rarely free or spontaneous but are founded on constructive, structural, or operational reasons. His nuanced organic forms are usually chained to a rigorous logical course that only constitutes the final link. This implies that the artistic creation of an architect does not have to contradict its scientific rigor. This is one of many reasons that prompted Norman Foster to assert that 'Gaudí's methods, a century later, continue to be revolutionary'. [7]

Fig. 4: Casa Batlló façade

Fig.5: The columns of the interior of Sagrada Família

2.2 Art Deco in Spain

Campi [8] investigated the channels through which Art Deco reached Spain and Catalonia. Working from a design history perspective she identified the industries that best represented it. She observed that the contribution of Spain to the Art Deco movement had been ignored by academic research in general and in particular as represented by the Art Deco 1910-1939 exhibition held at the Victoria & Albert Museum in 2003. [9] Despite this lack of official definition the style can be immediately recognized: it corresponds to a series of formal characteristics as follows:

- Schematic simplification and geometrization applied to both volumes and structural features and to decorative and ornamental features. This simplification of forms was inspired by avant-garde art (Cubism, Neo-Plasticism, Futurism, Constructivism, and rationalist architecture), setting out to emulate their formal repertoires rather than their social objectives.

- The evocation of the exotic. In an age of progress in communications, tourism, and scientific missions, this gave the design a cosmopolitan international look. The exotic ornamental motifs could be from geographically remote cultures, such as Africa or China, or from historically remote cultures, like Tutankhamun's Egypt or Meso-American cultures, or native traditions that had been well preserved, for example Maori culture in New Zealand or Flamenco folklore in Spain.

- The co-existence in the same object or interior space of rare and luxurious materials, generally polished and shiny (lacquers, varnishes, crocodile or snake skin, precious and semi-precious stones), with modern industrial materials (plastic, rubber, glass, aluminium, and chrome-plated metals).

- Western art is frequently characterised by its eclectic appropriation of historical styles. In the 1930s this took the form of a preference for classicism. During this time Art Deco took on a more rigorous and refined look. [9]

In the 1920s Art Deco was chiefly present in the luxury decorative arts and crafts, but in the 1930s it was also used to give the artefacts produced as a result of technological progress such as telephones, radio receivers, and electrical appliances a bold modern image. Before the First World War their designs had been primitive and frightening. In graphic art, advertising, and illustration, Art Deco was apparent equally in the predominance of simplification, geometrization, flat figures, and the use of exaggeratedly bold lettering reduced to simple circles, triangles and squares. [9]

As in the rest of Europe in Spain Art Deco coexisted with Modernism. This coexistence gave rise to heated controversies. There were many architects and designers who, upon noting the decline of Art Nouveau, promptly adopted Deco as a modern decorative style after the First World War. However, unlike the avant-garde, Art Deco was not trying to change the world. This aim was the preserve of Modernism, far more radical and politically committed in nature. Modernism set out to transform society and bring it into the twentieth century by collectivizing private life, freeing women from their domestic chores, and improving the health and hygiene of the body. This ambitious programme would be achieved using modern technologies, mass production, and the introduction of a mechanical and rational aesthetic, all in the context of the proletarian revolution. Although not all the architects of Modernism were openly left wing, it must be acknowledged that Modernism's programme was rooted in the ideas of equality and social justice as proposed by Socialism. [8]

2.3 Regionalism and Purism

In Spain regionalism and purism were stimulated by the idea of the 'regeneration' that Spanish intellectuals invoked after the crisis of 1898, through which it ought to be possible to harmonize technological progress with knowledge of the most profound Spanish identity. These rather nationalistic movements signalled the boom in rambling to explore the remotest places in the peninsula where it was thought the authentic Hispanic soul remained, uncontaminated. The romantic stereotype of Spain as a country capable of conserving its customs and traditions to a greater extent than neighbouring countries, which were being devoured by industrialization, provided creators with a measure of typically Art Deco exoticism. In the regionalist painting of Ignacio Zuloaga, Eduardo Chicarro, and Julio Romero de Torres themes appear of popular and purist Spain such as bullfighters, majas (young women), and flamenco dancers, which were a constant iconic visual element used by design, graphic illustration, and advertising of the time. As an example of this, Myrurgia the perfume manufacturer launched a series of new fragrances with images and names inspired by flamenco folklore and by the choreographic creations of the dancer Tórtola Valencia, an authentic muse of painters and writers of the period. [8]

According to Pérez Rojas, in Spain during the 1920s as well as these regionalist and purist tendencies there was an authentic rediscovery of the Baroque and the Renaissance that was expressed in architecture that he also considers Deco. It is true that Spanish cities were a long way from the great cultural capitals – Paris, London or New York – where Art Deco was created. Nevertheless, the most recent research has clearly shown that this style was spread all over the world through the channels of transmission typical of mass culture as no other had been before. Art Deco went around the world and reached countries far away from metropolitan centres such as India, China, Japan, South Africa, New Zealand, and Latin America. In all of these places Art Deco appropriated indigenous decorative cultures which it schematized, modernized, and turned into a contemporary style ready to be consumed by the middle and upper classes. Spain was no exception, and we have seen how quickly and eagerly it absorbed Art Deco through cinema, magazines, and exhibitions. In my opinion the studies of Art Deco in Spain ought not to be based on the idea that it was a provincial style, outmoded and peripheral, but they should explore how the strategies of schematic simplification, geometrization, modernization, luxury, mass consumption, and commercialization that define it were used to reinterpret local traditions. [8]

2.4 Ibero-American Exposition of 1929

A world's fair held in Seville from 9 May 1929 until 21 June 1930. The flags of the countries that attended this exposition can be seen in the poster shown in Figure 6; they are: Portugal, the United States, Brazil, Uruguay, Mexico, Peru, Argentina, Chile, the Republic of Colombia, Cuba, Venezuela, the Dominican Republic, Bolivia, Panama, El Salvador, Costa Rica, and Ecuador. Each Spanish region and each of the provinces of Andalusia were also represented.

Fig.6: Art Deco Poster of Ibero-American Exposition

Spain's Dictator, General Miguel Primo de Rivera, gave the opening address, and King Alfonso XIII officially opened the exposition. The purpose of the exposition was to promote tourism and to improve relations between Spain and the countries in attendance, all of which have historical ties with Spain either through colonization (parts of Spanish America and the United States) or by political union (Portugal and its former colony Brazil). Other countries were represented at the International Section in Barcelona. The exposition was smaller in scale than the International Exposition held in Barcelona during that same year, but it was a significant representation of regional design styles. The city of Seville had prepared for the Exposition for the previous nineteen years. The exhibition buildings were constructed in María Luisa Park along the Guadalquivir River. Most of the buildings were built to remain after the closing of the exposition. [10]

2.5 Retro-futurism

Blending historical styles with futuristic technology, retro-futurism explores the themes of tension between past and future, and between the alienating and empowering effects of technology. It is a trend in the creative arts evidencing the influence of depictions of the future produced prior to 1960. Initially established in artistic creations, retro-futurism has extended into the disciplines of fashion, architecture, literature, and film. It consists of two related perspectives, that are, the future as seen from the past, and the past as seen from the future. The former trend is directly inspired by the imagined future which existed in the minds of the creatives of the pre-1960 period who attempted to predict the future, either through projections of existing technology or in science-fiction stories. These futuristic visions are reconstructed and updated for the present, and offer a nostalgic, counterfactual image of what the future might have been, referred to as retrofuturism proper. The later trend is the inverse of the first. It is called futuristic-retro. It starts with the retro appeal of old styles of art, clothing, etc., and then incorporates modern or futuristic technologies, creating a blend of past, present, and future elements. [11]

Retro-futurism proper is inevitably influenced by the scientific, technological, and social awareness of the present, while futuristic-retro creations are more than just copies of their pre-1960 inspirations. Instead, they are given a new, ironic twist from a modern perspective. The retro-futuristic trends in themselves are not time-specific. For the time period of a narrative, it might be a 'what if' present based on a unique technology, an outlandish version of the future, or an alternative past in which fictitious or projected inventions of the past were a realisable technology. While retro-futurism, due to the varying time-periods and futuristic visions to which it alludes, does not provide a unified thematic purpose or experience, a common thread is dissatisfaction or discomfort with the present, to which retro-futurism provides a nostalgic contrast. A similar theme is dissatisfaction with the modern world itself. A world of high-speed air transport, computers, and space stations is (by any past standard) 'futuristic', yet the search for alternative and perhaps more promising futures suggests a feeling that the desired or expected future has failed to materialize. Retro-futurism suggests an alternative path and, in addition to pure nostalgia, may act as a reminder of older but now forgotten ideals. [11]

The Spanish design brand Sancal has created a line of futuristic-retro furniture. The 'Futura' collection, shown in Fig. 7, has a muted aesthetic, primarily using white, cream, brown, mustard yellow, and silver colours. It features couches and chairs with rounded features and there are large circular cut-outs in the walls. Thematically the exhibition has the emotional sense of an abandoned spaceship. [12]

Fig. 7: Sancal futuristic-retro furniture

3. ADDITIVE MANUFACTURING AND DESIGN

Additive Manufacturing (AM) is reaching many different industries and markets due to the increase in performance and reliability of the industrial process and the improvements in the materials to be printed. Beside traditional sectors like aerospace, automotive, or medical industries, AM is every day more present in other industries, like consumer products, fashion, maritime, or even furniture. [13]

As the technology consolidates, the knowledge needed to apply this technique correctly is also wide spreading [14]. This is not only due to the specific training or education of engineers and industrial designers towards the use of this technology, which is necessary and is every day more accessible, but also because of the rise in awareness of what this technology means and what it can offer to product design and development. [15][16]

The increase of works, publications, and success cases for different industries, together with specific workflows and novel software tools for Computer Aided Design (CAD), engineering (CAE), topology optimisation (TO), slicers, etc.) is generating the framework needed to make industrial designers step forward and start considering this process as an option when developing products. The consolidation of this technology offers industrial designers opportunities to receive specific training in Design for Additive Manufacturing (DfAM) and in a better knowledge of the technology. [17]

Design for Additive Manufacturing must be integrated into the traditional design process right from the beginning. Structured design processes, such as those proposed for industrial design engineers all around Europe like Delft design methodology, VDI, etc., [18] should expand, integrating a new industrial process that has specific advantages and will happily sit alongside all the other traditional industrial processes. The main advantages are the possibility of building highly complex shapes, or the capability of changing every printed part without needing to adapt or generate new tools, the possibility of combining different materials and controlling specific properties at different locations of the printed parts, like density of centre of mass. [19][20][21][22] In addition, it is important to know the drawbacks of this process, like the lack of dimensional accuracy, the anisotropy of the properties of the printed parts, and long manufacturing times, together with size and cost. [23][24][25]

Understanding both the advantages and the limitations will give industrial designers the opportunity to correctly select the technology and apply DfAM for product development adding value to the projects. From the authors' point of view small productions units, with really complex shapes and with a high degree of customization, is the place where this technology should be applied to fully exploit its potential.

There are many reasons why the application of this technology really matches the goal of this research into interior design for luxury yachts such as the proposed concept of La Transatlántica. Luxury products always demand the use of the latest breakthroughs in technology such as AM. Furthermore, the really complex shapes that can only be achieved through this technological process can ensure that the exclusivity, the customised parts, and the tailored design that luxury markets demand will be realised. Given these considerations, designing furniture pieces for interior design in luxury superyachts is fully justified and, in the next sections, a study case about this is presented.

The starting point of this design was to explore the possibilities of AM to generate new furniture concepts with a specific design brief: to fully equip a flamenco performance venue in La Transatlántica superliner design concept. The briefing pointed out that a retro futurism concept with some influences drawn from Art Nouveau and from Andalusian culture should address the design proposals. Before describing the design process a benchmarking of existing similar products has been undertaken to inform the product development process. In Figure 8 you can see a selection of 3D printed furniture.

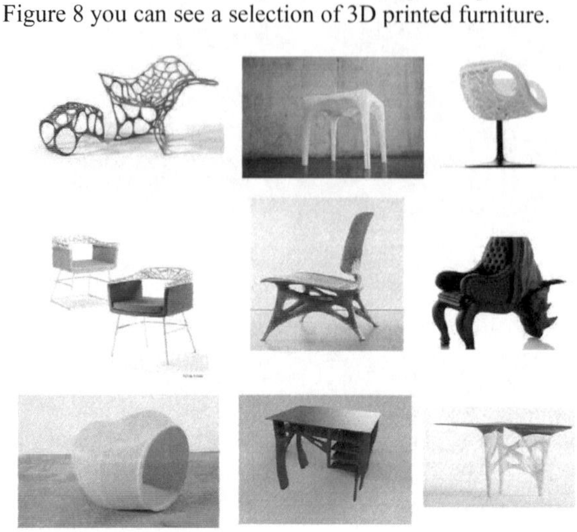

Figure 8: Different furniture pieces developed with AM technologies

When designing a new product, it is necessary to follow a complete design process that takes account of different studies such as trends, market research, user research, ergonomics, technology, etc., when aiming for competitive results. [18] In this case the authors, being aware of their limitations, have only focused on the application of novel design tools for additive manufacturing for the development of new concepts with really specific geometrical characteristics that create a shape study or styling concept more than a fully analysed design solution.

4.0 DESIGN-DRIVEN INNOVATION PROCESS

The process of Design-Driven Innovation involved the development of a design scenario through engaging in unstructured discourse with a range of interpreters in technology and cultural production. In this project this included: superyacht design; interior design; product design; additive manufacturing and design; and naval architecture. The design process of the flamenco venue on a transatlantic superliner began with a significant amount of cultural research, including benchmarking of key flamenco venues in Andalucía, as well as key regional architectural and interior styles.

4.1 Exterior form and GA development

The second stage involved the development of a conceptual exterior form based on retro-futurism and a parametric GA to contextualise the design of the Flamenco venue into a holistic vessel proposal. The exterior form was developed first in side profile using the deck-line heights of the previous EBDIG 280m pentamaran superliner project [1] and then in front ¾ view over a front ¾ underlay of the same vessel. The key objective of the exterior design was to allude to the design heritage of the golden era of superliners such as the Normandie by developing a sleek, streamlined, and dynamic shape. This should convey a sense of power, presence, and capability of the innovative high speed pentamaran platform. In addition, it should emphasize a sense of latent capability of speed when stationary, as well as developing an exterior form with retro-futuristic design communicating the capability and technology of the platform as a form language relating all the visual components. To communicate the speed and modernity of the design visual form techniques were used to reduce the perceived visual mass of the superstructure and to accentuate the perceived length of the vessel.

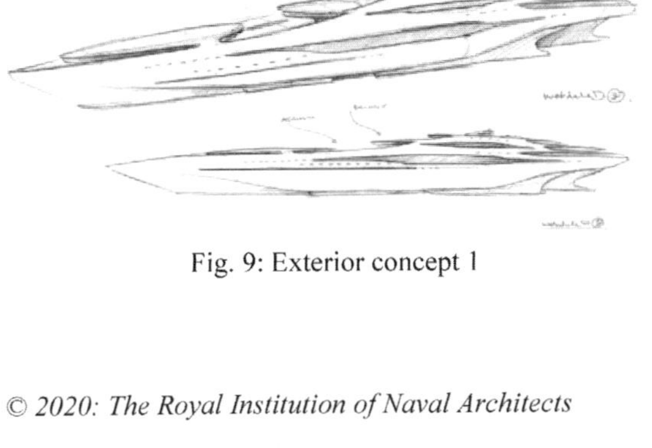

Fig. 9: Exterior concept 1

Exterior concept 1, shown in Fig.9, has the dynamic and organised sense of the form connecting the large window areas through similarity of form. In addition, the window areas visually connect to the two, peak, cap-like surfaces fore-facing on the superstructure at mid-ship. The sculpted fore deck volume of the flamenco venue connects to the shearline on both port and starboard. In doing so it creates a courtyard bounded by the rear of the flamenco venue and the rest of the superstructure aft of it, with flowing bulwarks on either side.

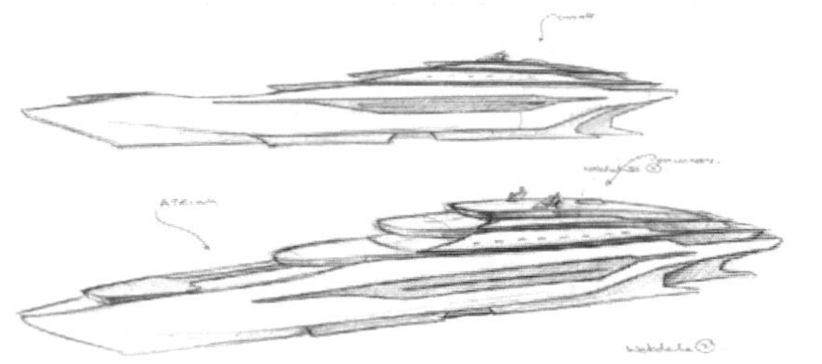

Fig. 10: Exterior concept 2

Exterior concept 2, shown in Fig. 10 above, has the rectilinear form sensibility of Art Deco combined with the use of streamlining to accentuate the elegant form. It has a linear surface splitting band in the middle of the DLO to accentuate length through the perceived proportions of the split areas of glazing. The bow has a fore deck open space with the flamenco venue rising out of the deck at the bow area.

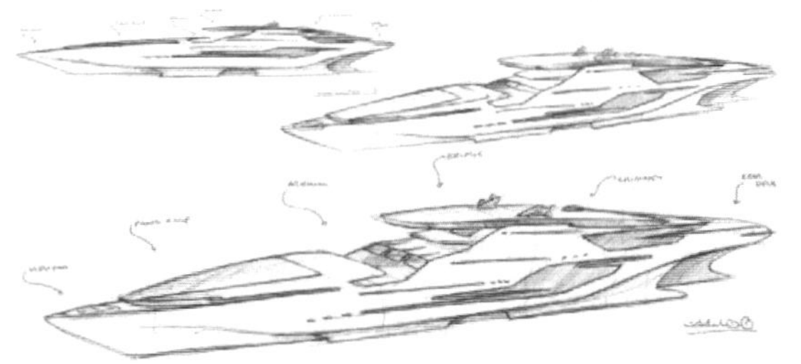

Fig. 11 Exterior concept 3

Exterior concept 3, shown in Fig.11 above, achieves the design language of a spaceship through the implementation of retro-futuristic styling. This is achieved through the use of thin rectilinear DLOs which are visually extended by the use of square portholes. These are used to balance the visual mass of the large geometry glazed areas. The large shuttle craft styled volume on the fore deck connects to the shearline on both the port and starboard. It creates a courtyard type of volume between the aft of it and the rear super structure which has balconies overhanging the courtyard to reinforce this architectural meaning of the space. This was the chosen exterior design direction: a Photoshop render of it is shown in Fig.12 below where the sleekness of the form is emphasised by boat tailing the rear of the vessel and sculpting the aft surfaces on the upper decks to visually connect with the feature on the main deck, namely the flamenco venue. This acts as a focal point in the bow area. The thin lipped streamed-lined surface at the top of the superstructure adds emphasis to the courtyard volume as an absence of space, making it an architectural focal point of the design. Combined with the aft position of the superstructure's visual mass this produces a dynamic and slightly aggressive imposing stance. The shape of the long horizontal windows on each deck interacts with the visual proportions of the deck to accentuate the visual length of the vessel.

Fig. 12: Rendered final concept

Once the final form was decided a parametric GA proposal was developed by interrogating the side profile which had deck-line heights and the front ¾ render to estimate the profile shapes for each deck of the GA. The GA layout shown in Figure 13 was then developed using people flow analysis with user scenarios at different times of day. Here Deck 2 shows the cargo hold designed for specialist TEU loading and the engine room. The tank deck is below Deck 2. Deck 3 has crew accommodation and storage aft of the crash bulkhead; this leads to a mezzanine music hall which has stairs and elevators to connect to the balcony on the above floor. At mid-ship of deck 3 there is a restaurant and connected galley in the spirit of the Normandie. It has a light tunnel which passes up to the courtyard space three decks above. The restaurant is surrounded on either side by guest suites. Aft of the restaurant is a full beam exhibition space, aft of which there is a cinema flanked on either side by lounge spaces. Aft of the cinema is a flamenco bar, flanked by guest suites. Aft of it are the lifts and a hospital, behind which are more guest suites and behind then crew accommodation.

Deck 4 has storage and galleys aft of the crash bulkhead. Aft of this there is a lounge area leading to a 3-deck high atrium. Aft of the atrium is a tapas bar leading to a larger atrium, which is two decks high and which leads onto an art gallery, which in turn leads on to another atrium, which is four decks high and which leads into the cinema. Both art gallery and atrium are flanked by guest suites. Aft of the cinema is the balcony of the flamenco bar, which is surrounded by guest suites. Deck 5 has the lower bar of the flamenco venue aft of the crash bulkhead, followed by guest suites, a full beam lounge area, and then an atrium. Aft of the atrium are guest suites followed by a lounge space and an atrium. Aft of this are the lifeboats, more guest suites, a spa, and more guest suites leading to the swimming pool on the transom.

Deck 6 has the balcony area of the flamenco bar at the bow leading to an atrium followed by a tapas bar which leads onto the courtyard space. Aft of the courtyard are two lounge spaces leading on to an atrium and elevator which are surrounded by guest suites. Fore of the outside swimming pool there is a large panoramic restaurant which stretches over two decks. Deck 7 has panoramic lounge followed by guest suites followed by a restaurant, with elevators and stairs between the lounge and guest suites. Finally, deck 8 has the bridge, crew accommodation, and a panoramic restaurant.

4.2 Interior Design Theme and Methodology

The approach adopted here was the development of a retro-futuristic theme for the flamenco venue as a juxtaposition of Spanish Art Deco and Modernisme, as well as embracing the historical cultural connection of La Trasatlántica (CTE) having terminals in Cadiz and Barcelona at different times. The design realisation of this theme was informed by the creative potential of 3D printing and biomimetic forms that can be achieved through the associated specialist CAD tools. Significant research was carried out into both contemporary and historical Spanish interior design. This generated a range of mood boards for several themes. The key interior areas were developed and evaluated in the context of the millennial persona demographics. The flamenco performance room requirements were informed by benchmarking key flamenco establishments across Andalucía. Connections between the vessel's exterior and the interior was achieved through the commonality of the retro-futuristic theme.

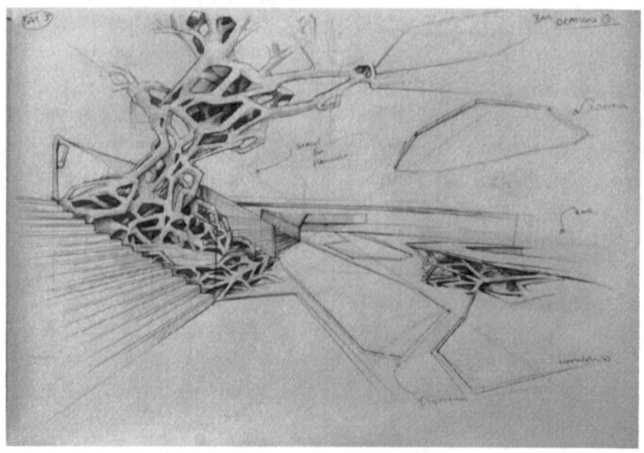

Fig 13: Biomimetic tree staircase feature -side view

The retro-futuristic theme of the flamenco venue as a juxtaposition of Spanish Art Deco and Modernisme was explored through sketching and informed by research and the DDI discourse. A parametric spatial layout of a mezzanine with balcony was identified as the best solution. This required a focal point to emphasize the transition between the nature of the two spaces where the lower is a performance space and the upper is a social lounge area. Explorative forms of a biomimetic tree and column of light feature were evaluated, as shown in figures 13 to 16. It was decided that the theme of the interior design should be Art Deco and the theme of the furniture Modernisme. The interpretation of Art Deco within a retro-futuristic theme for the bar is shown in Figure 17. Photoshop renders of the Art Deco influenced features are shown in Figure 18. While parametric biomimetic furniture forms were evaluated through sketching as shown in Figure 19, it was realised that a furniture CAD methodology would be required to inform the design process with the capabilities and limitations of 3D printing and software based in cellular design or Lattices.

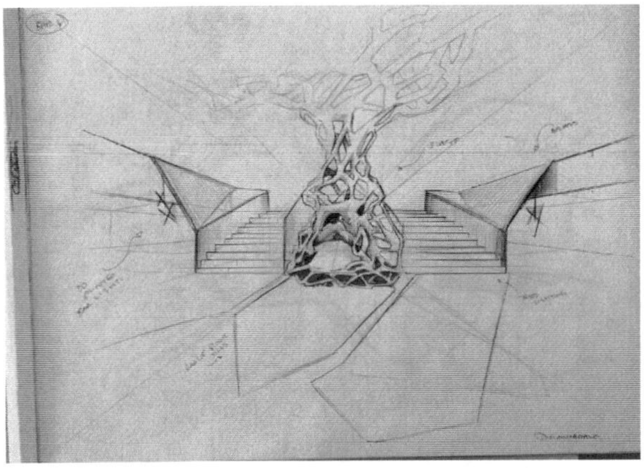

Fig 14: Biomimetic tree staircase feature -front view

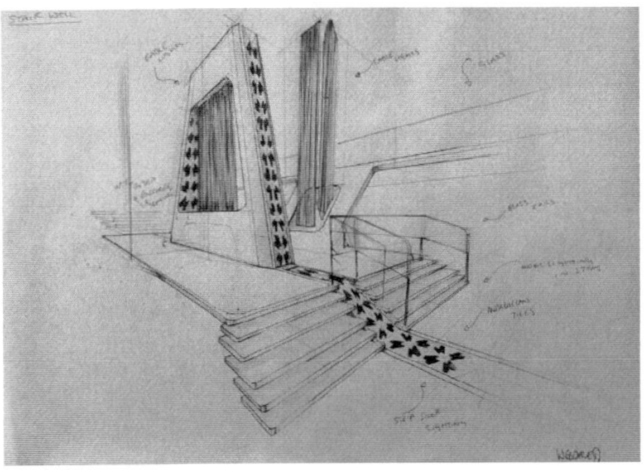

Fig.15: Column of light architectural feature

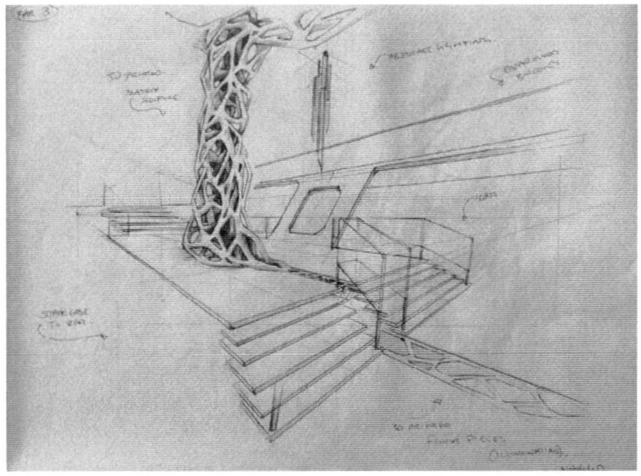

Fig.16: Biomimetic tree landing space feature

Fig. 17: Art Deco influenced bar with lighting feature

Fig 18: Renders of Art Deco influenced features

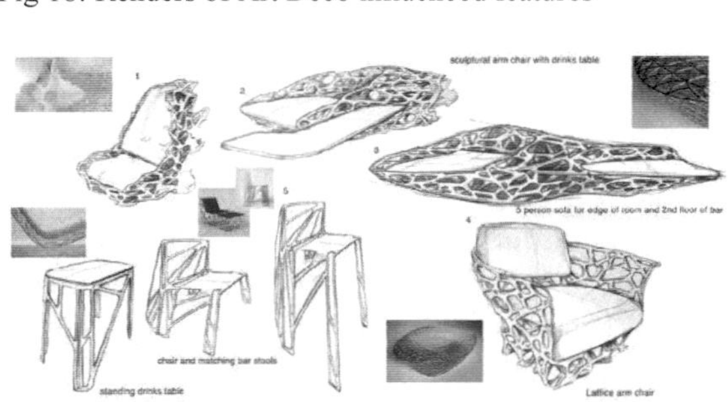

Fig 19: Biomimetic furniture sketch development

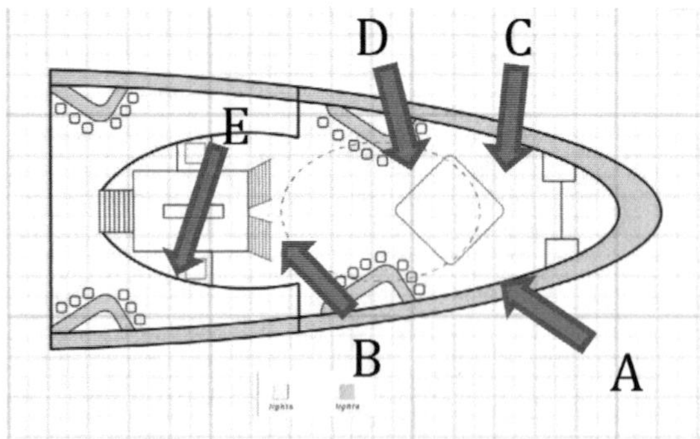

Fig. 20: GA of flamenco venue

The GA of the flamenco venue located at the bow area is shown in Fig.20. The view from camera location A is shown in Figure 21; this is split into a two-tier configuration. The stage can be seen in the foreground, flanked on either side by bars, behind which there is the tableau leading to an atrium style staircase.

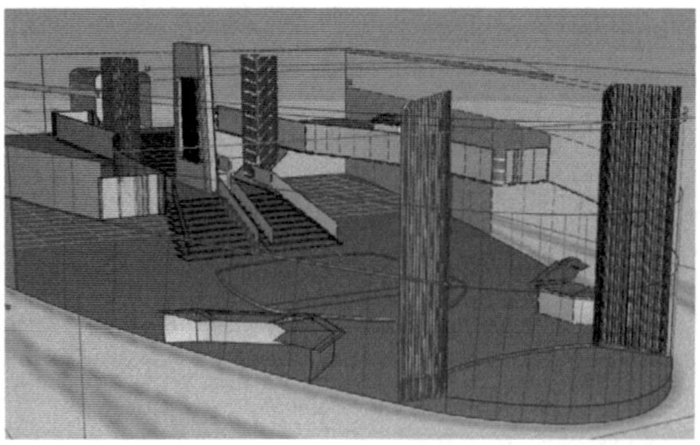

Fig. 21: Flamenco venue view from camera location A

Fig 22: Flamenco venue view from camera location B

The visually floating staircase appears to be suspended by structural light cables and a column of light architectural feature acting as a focal point, as shown in Figure 22. The purpose of the column of light feature acting as a focal point is to emphasize the transition between the nature of the two spaces: the lower as a performance space and the upper as a social lounge area. When standing in the balcony area, shown in Figure 23, the column of light feature draws your eye down the stairs to the tableau and bar area set underneath the chandelier, which itself makes the tableau a focal point through symmetry. This natural read of the space choreographed by light encourages users down to the performance area.

Fig 23: Flamenco venue view from camera location C

Fig 24: Flamenco venue view from camera location C

Fig 25: Flamenco venue view from camera location D

The performance space has a large chandelier which connects the two floors together by virtue of having symmetry with the tableau. In turn this makes the tableau into an architectural focal point. This is shown from the view side of the stage in Figure 24, and the view from the bar right of the tableau in shown in Figure 25. The chandelier is hung from the ceiling by light bars. These light bars are then connected to light plates. The light plates are inspired by the geometric patterns specific to Andalucia that date from the Moorish Empire. The design intent is for these light cables to pulse with a warm glow to engage the users in an emotionally sustainable experience, similar to that of the Apple Mac heartbeat light.

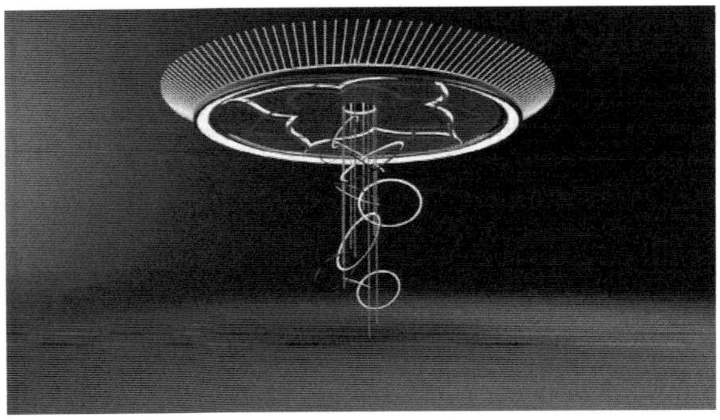

Fig 26: Small balcony chandelier

Hanging from the surrounding balconies are two smaller chandeliers which connect to the larger chandelier by implementing the suspended element design theme as shown in Figure 26 above.

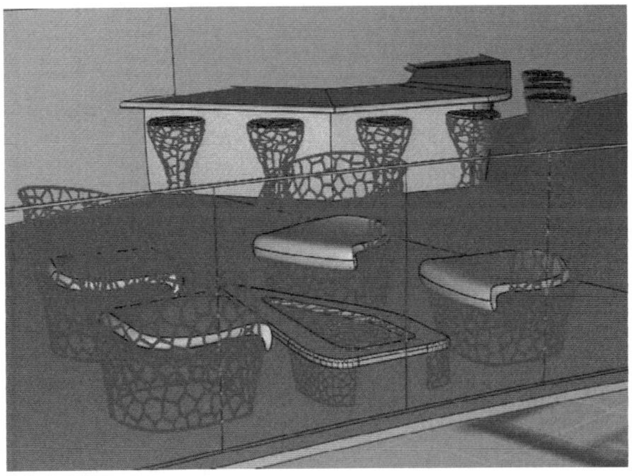

Fig. 27: Balcony bar from camera location E

On both floors four small bars have been designed looking out on to an open space holding the stage and tableau floor used for flamenco performance. The shape of the room is oval to enhance the acoustics. The balconies overhang the lower bars so passengers on both floors can observe the band and dancing. The second floor will hold more 3D printed tub chairs, sofas and tables, as shown in Figure 27, allowing more space on the bottom floor is to encourage flamenco dancing.

4.3 Furniture CAD Methodology

Before describing the design process the novel design tools will be introduced. We will address the design using a design software based in cellular design or Lattices. The software used for this task is called Ntopology™ and allows the user to transform different geometries such as surfaces or solid bodies in different lattices structures predefined or defined by the user. The main advantage of these tools is that they make possible the reduction in the mass of the developed parts maintaining the mechanical performance. This specific geometry also involves a better performance of the parts in specific engineering needs like heat dissipation or cooling with complex fluid channels designs. [26]

In the study case here presented to illustrate the process employed different pieces of furniture were selected to apply this novel design tools to. The selected items were a stool, a chair, and a table. With all the items the first task was to analyse which parts of every piece of furniture could be optimized and transformed in lattices. The parts that could be transformed were called design spaces and the ones that should not be changed due to functional restrictions were called no design spaces. The initial geometries or parts were designed in an assembly mode in Alias CAD software from Autodesk™, a specialized design tool predominantly used in automotive design and industrial design for generating Class A surfaces using Bezier surface and NURBS modelling method. Once the design was completed all different parts were exported in a STEP (Standard for the Exchange of Product model data) format defined in the standard ISO 10303. These files are imported to Ntopology Element platform, an academic version. The newest and more extended platform is called Ntop; this includes a design, topology optimisation, and structural analysis module.

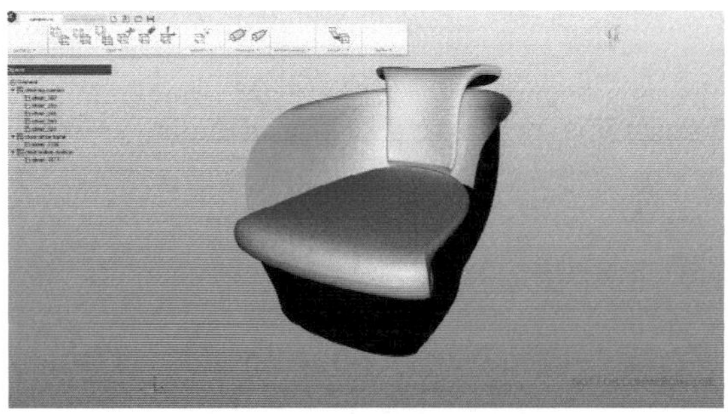

Fig 28: Imported CAD surfaces

Once the parts are imported as shown in Figure 28 the design spaces are selected, and a specific lattice structure is defined and applied to them, as shown in Figure 29. In this case a stochastic surface lattice is selected as the design spaces are defined to be surfaces, for the chair, stool, and table. In all cases a detailed modification of the proposed lattice is required to obtain continuity between the design and no design spaces, and to adapt the design to the manufacturing process in what can be consider a DFM (Design For Manufacturing approach). The next step is to provide thickness to the defined lattices; the thickness considerations attend to the selected manufacturing process depending on its accuracy. Finally, the optimized parts are meshed and exported in STL files ready to define the process characteristics and to generate the machine code for printing (GCODE).

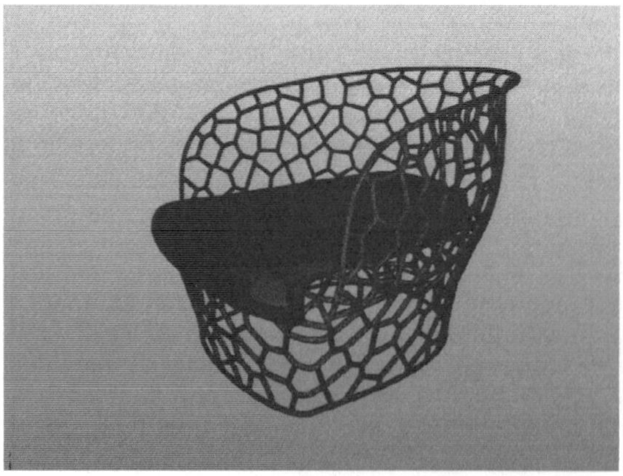

Fig 29: Application of lattice structure

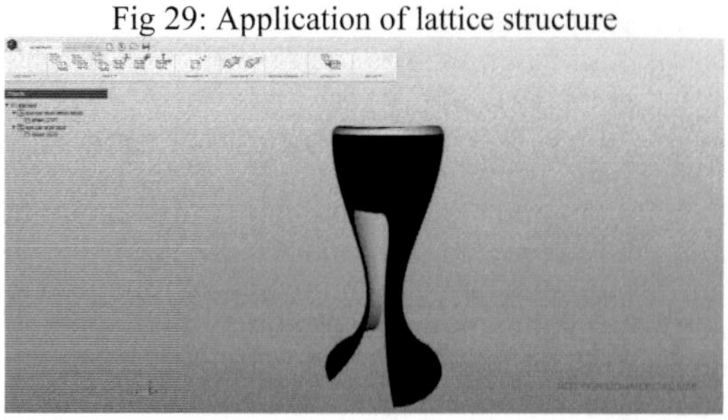

Fig 30: Design surfaces

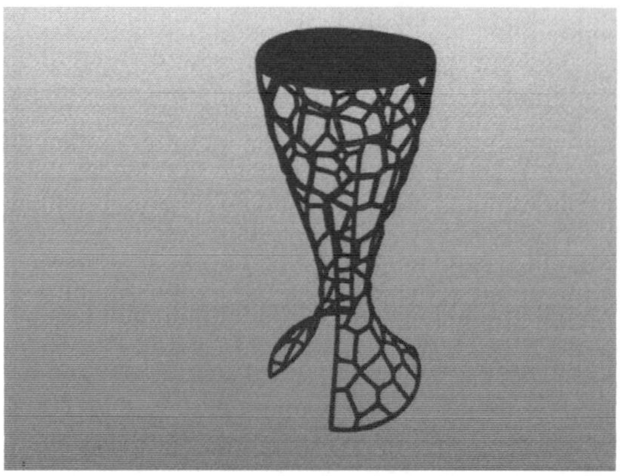

Fig 31: Lattice structure of design surface

Ntopology screen captures are shown above. Figure 30 shows design and no design spaces of the furniture pieces to be optimized. Figure 31 shows thickened lattice and meshed design ready to be exported. Due to the size requirements of the selected parts in this case study large format additive manufacture (LFAM) in a one cubic meter equipment is defined. This equipment belongs to the Material extrusion processes defined in the AMST F42 committee for AM, [27] and could be fed with polymeric thermoplastic material in the shape of filaments or pellets. In this case two scaled down prototypes have been printed using a Creality desktop printer CR10 5S using PLA material to validate the design concepts as shown in Figure 32.

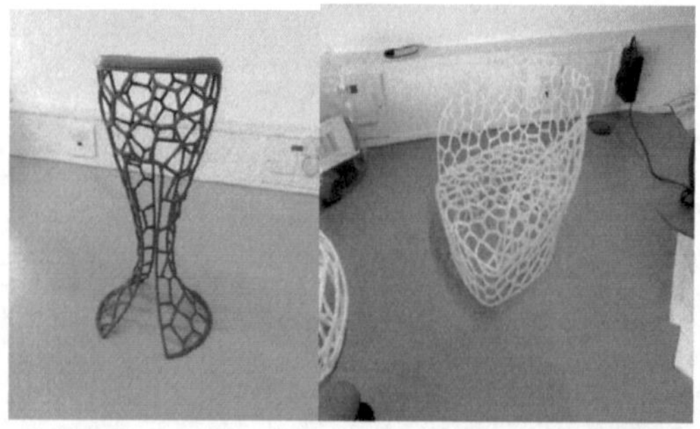

Figure 32: 3D printed prototypes of the chair and stool

The geometrical characteristics of the design (single wall lattice) and the anisotropy of the process makes the Z loading direction the weakest due to the interlayer bonding. This suggests that the printed parts should be used as neat shape models for aluminium sandcasting as shown in Figure 33's rendered images. The results will be interesting organic random shapes that give the developed products the exclusivity and customization defined in the design brief.

Figure 33. Rendered images of final furniture concepts

5. DISCUSSION

This Design-Driven Innovation approach to developing a retro-futuristic style of a flamenco venue for a transatlantic superliner as a homage to CTE was an explorative journey. The context of the cultural specificity of the design was supported by a cultural immersion of a placement student from Coventry University at UCA in Cadiz with direction feedback from researchers operating within their own culture. The flamenco performance room was informed by benchmarking a number of key flamenco establishments across Andalucía.

The use of innovative lighting technology constituted a reinterpretation of the innovation that lighting was and how it was used as a design element in Art Deco. Figure 34 below shows design for the foyer of the Strand Palace Hotel by Oliver Bernard between 1930 and 1931. Bernard began his career working as a stage designer in Britain and the US and was technical director of the British Exhibition in Paris in 1925. His practice was informed by this experience. With its mirrored glass entrance and interiors the Strand Palace brought Hollywood to London. Art Deco in Britain was in the French Style: it glamorised innovation and creativity. [9] The retrofuturistc approach of the stairwell reflects this design meaning.

The overall design objective for the superliner was to extend the proposal of the previous project into a new cultural context, to examine the narrative of cultural diversity within the Spanish diaspora as a reenvisioning of the objectives of the Ibero-American Exposition. In further work the other interior areas of the GA will be developed and refined to truly embrace the cultural diversity within the Spanish diaspora as a reenvisioning of the Ibero-American Exposition. There will be a key focus on the ports of the old routes of Trasatlántica (CTE) such as Cuba.

Fig. 34: Foyer of the Strand Palace Hotel [9]

Campi [8] perceives the neo-Baroque buildings of the 1920s as examples of terminal historicism rather than the beginnings of a modern idiom. She did not find any schematic and stylized interpretations of the Baroque in Spanish furniture such as the ones made by the great French ensembliers Jacques Émile Ruhlmann and Süe et Mare. She also objected to the use of the adjective 'deco' to describe revivals of the Plateresque style, as she found it old-fashioned, anachronistic, and dull. Accordingly, transferring it to a style of furniture resulted in the 'Spanish remorse' style, examples of which decorated aristocratic mansions. In her opinion, the idea that the Baroque and Renaissance revivals were national styles reinterpreted in a modern way is questionable because they appear to be nostalgic evocations of a glorious past that gave interiors a gloomy old-fashioned look. The cultural connection of Trasatlántica to both Cadiz and Barcelona provided the inspiration for the juxtaposition between Andalusian Art Deco and Catalan Modernism within the retro-futuristic design theme to create a positive vision of the future drawn from the past exceeding the limitations of the Neo-Baroque style identified by Campi. [8]

The design workflow as an exploration of process and the link between concept and realisation through printing 1/5 scale models has identified the need to have a more detailed design workflow to examine the potential of lighting as a key design element which can add a high degree of emotion to the user experience. The workflow here was based on sketching ideation informed by visual reference and a depth of cultural research. Photoshop was then used to examine colour, and CAD was used to examine the influence of lighting elements on top of an intuitive interior design capability. The making aspect of the model furniture helped to develop a spatial analysis. In further work the use of CAD rendering and full-size models will enable the visual characteristics of embedded lighting technology to be calibrated for a range of translucent and transparent 3D printing materials. This will facilitate a more efficient and effective workflow whereby a design can be fully developed using CAD rendering and VR through systems such as Oculus Rift to enable the printed artefact to behave as designed in a simulated VR environment.

The engagement with sustainable luxury is partially addressed in the previous DDI proposal with the use of cargo. This could be further developed by the introduction of fuel cell technology and hydrogen high pressure storage systems at 750bar. Currently Nedstack have developed 1MW TEU based fuel cell units for industrial applications but given that the power requirements would be in excess of 75mw this would equate to 75 TEU in addition to a significant volume of hydrogen high pressure gas cylinders. This will be evaluated in further work in terms of fuel cell technology innovation and the associated vessel architecture required. Recently DNV-GL identified hydrogen as the ultimate zero-emission solution for shipping. Ulstein and Nedstack have developed the first complete hydrogen-fuelled vessel, the Ulstein SX190 Zero Emission DP2 construction support vessel. Sea trials of a newbuild Ulstein SX190 Zero Emission are planned for 2022. The PEM fuel cells used are fuelled by hydrogen from containerized pressure vessels. These hydrogen storage containers can be loaded and unloaded using normal container handling equipment, thus eliminating the need for expensive bunkering infrastructure and providing worldwide operational flexibility. [28]

In terms of engaging in a sustainable luxury agenda with 3D printed bespoke furniture completely biodegradable bioplastics will be required. The US company Algix is developing bioplastics from algae. The company harvests algal biomass from waters with algae blooms and uses it to produce biodegradable filaments for 3D printing. The removal of algae biomass ensures more stable ecosystems and is intended to reduce the risk of a recurrence of algae bloom. The biodegradable filaments for 3D printing consist of about 20% algae components and 80% PLA, a biopolymer produced by polymerising lactic acid, which in turn is obtained from plant sugars. An additional benefit of algae-based filaments is that they consume less energy during printing than conventional filaments because they can be processed at a lower temperature. [29]

6. CONCLUSIONS

Through retro-futurism there is an opportunity for a design renaissance of Spanish Art Deco which has never had the international cultural recognition that it deserves. The superliner concept as a design artefact is an opportunity to re-examine the narrative of cultural diversity within the Spanish diaspora as a reenvisioning of the objectives of the Ibero-American Exposition. It will be developed further as a collaborative project within the EBDIG network.

The design workflow as an exploration of process and the link between concept and realisation through printing 1/5th scale models has identified the need to have a more detailed design workflow to examine the potential of lighting as a key design element which can add a high degree of emotion to the user experience. The workflow here was based on sketching ideation informed by visual reference and a depth of cultural research. Photoshop was then used to examine colour and CAD was used to examine the influence of lighting elements on top of an intuitive interior design capability. The making aspect of the model furniture helped to develop a spatial awareness as part of the design process.

7. ACKNOWLEDGEMENTS

The design work presented in the paper was carried out by Dylan Webdale, a Boat Design student studying Automotive & Transport Design at Coventry University, while on an Erasmus+ funded work placement at the Industrial Design Manufacturing Research Centre at Cadiz University. The renders shown in Fig.22-24, were produced by Jonathan Elcock-jules, a Boat Design student in the Professional Enhancement Year of the Automotive & Transport Design Course at Coventry University. The authors wish to thank Dr Antonio de Querol Sahagun from Cadiz University's Department of Naval Architecture, for his significant level of support relating to the technical discussions of the vessel platform.

8. REFERENCES

[1] MCCARTAN, S., VERHEIJDEN, R., and CROTTY, P., 'Design-Driven Innovation: Next generation High Speed Transatlantic Superliner', The Royal Institution of Naval Architects, Conference: Marine Design, Coventry, UK, 3-4 September 2014

[2] VERGANTI, R. ,'Design-driven innovation: changing the rules by radically innovating what things mean', Harvard Business School Publishing Corporation, ISBN 978-1422124826, 2009

[3] https://www.histarmar.com.ar/LineasPaxaSA/56CiaTrasatlantica.htm?fbclid=IwAR15iWY1cR2tApneb52DC2zOSS5X27IzTkoga5NqPugPjcaKg-dn2tjFlf0 accessed 01/01/2020

[4] R. Carr, "Modern Spain, 1875-1980", ISBN 9780192801296, Oxford University Press, 2001.

[5] S. Holguín, "Flamenco Nation: The Construction of Spanish National Identity," ISBN 9780299321802, University of Wisconsin Pres, 2019.

[6] J. J. Lahuerta, A, Gaudí, "Antoni Gaudi, 1852-1926: Architecture, Ideology, and Politics," ISBN 9781904313205, Electaarchitecture, 2003.

[7] C. SALAS MIRAT, C. BEDOYA FRUTOS, and J.M. ADELL ARGILÉS, "ANTONIO GAUDÍ, PRECURSOR DE LA SOSTENIBILIDAD Y LA BIOMIMÉTICA EN LA ARQUITECTURA, CON 100 AÑOS DE ANTELACIÓN," ACE: Architecture, City and Environment, 13 (37): 71-98, 2018. DOI: http://dx.doi.org/10.5821/ace.13.37.5348

[8] I. Campi, "Art Deco in Spain: Mass Culture and Style". In Anna Calvera (ed): *Barcelona Design System (1914-2014) Essays on Local History*, Barcelona, GRACMON/ Edicions de la Universitat de Barcelona, 2018.

[9] C. Benton, T. Benton, and G. Wood, "Art Deco 1910-1939", Harry N. Abrams, 2015. ISBN 9781851778331.

[10] J. Mendelson, "Documenting Spain: Artists, Exhibition Culture, and the Modern Nation, 1929-1939", Penn State Press. ISBN 9780271047201.

[11] https://design-engine.com/retrofuturism-the-tomorrow-of-yesterday-or-the-past-rewritten/

[12] https://www.trendhunter.com/trends/retrofuturistic-furniture

[13] I. Gibson, D. W. Rosen, B. Stucker, and and B. S. I. Gibson, D. W. Rosen, Additive Manufacturing Technologies: Rapid Prototyping to Direct Digital Manufacturing, vol. 54. Springer, 2009.

[14] ASTM-F42.91 Subcommittee, "Standard terminology for additive manufacturing technologies," ASTM Des. F2792-12a, Vol.10.04., 2015.

[15] F. Abad, V. Casal, D. Moreno, and S. I. Molina, "Fabricación Aditiva en la Construcción Naval en el contexto de la industria 4.0," Vol 1, pp. 33–46, 2017.

[16] D. Moreno, V. Casal, and S. Ignacio, "Large-format polymeric pellet-based additive

manufacturing for the naval industry," Addit. Manuf., vol. 23, pp. 79–85, 2018.

[17] www.ceadgroup.com.accessed 3rd December 2019.

[18] Lonnie J. Love, "Cincinnati Big Area Additive Manufacturing (BAAM)," 2015.

[19] L. J. Love et al., "Breaking barriers in polymer additive manufacturing," in International SAMPE Technical Conference, 2015.

[20] B. K. Post, R. F. Lind, P. D. Lloyd, V. Kunc, J. M. Linhal, and L. J. Love, "The Economics of Big Area Additive Manufacturing," Proc. solid Free. Fabr. Symp., pp. 1176–1182, 2016.

[21] P. Parandoush and D. Lin, "A review on additive manufacturing of polymer-fiber composites," Composite Structures, vol. 182. Elsevier Ltd, pp. 36–53, 15-Dec-2017.

[22] T. Hofstätter, D. B. Pedersen, G. Tosello, and H. N. Hansen, "State-of-the-art of fiber-reinforced polymers in additive manufacturing technologies," J. Reinf. Plast. Compos., vol. 36, no. 15, pp. 1061–1073, 2017.

[23] E. Yasa and K. Ersoy, "Additive Manufacturing of Polymer Matrix Composites," Aircr. Technol., 2018.

[24] http://www.thermwood.com/. Last accessed 3rd December 2019.

[25] L. J. Love, "Wide and High Additive Manufacturing (WHAM)" 2017.

[26] Https://www.cjrprop.com/manufacturing/. Last accessed 3rd December 2019.

[27] P. C. Chesser, R. F. Lind, B. K. Post, A. Roschli, L. J. Love, and K. T. Gaul, "Using Post-Tensioning in Large-Scale Additive Parts for Load Bearing Structures," 29th Int. Solid Free. Fabr. Symp., 2018.

[28] https://www.offshoreenergytoday.com/ulstein-develops-its-first-hydrogen-powered-offshore-vessel-design/ accessed 01/01/2020

[29] https://en.reset.org/blog/does-future-3d-printing-lie-bioplastic-made-algae-07092018 accessed 01/01/2020

8. AUTHORS' BIOGRAPHY

Dr Seán McCartan is Course Tutor, Automotive & Transport Design, at Coventry University, UK. He is the Editor of the RINA journal *The International Journal of Marine Design.* His key research area is TOI (Transfer of Innovation) from other sectors to the marine industry, in the areas of Design-Driven Innovation (DDI), advanced visualisation, and Human Systems Integration (HSI). He leads the EBDIG (European Boat Design Innovation Group) network. This includes Chalmers University, Genoa University, TU-Delft, and a number of leading European marine design consultancies. He was project co-ordinator for the Leonardo TOI project EBDIG-WFSV (European Boat Design Innovation Group - Wind Farm Support Vessels), which aims to develop online training material for naval architects in Human Factors; WFSV design (Industrial Design); and WFSV mothership design (Industrial Design).

Daniel Moreno Nieto holds a PhD in material science and product development, having presented his doctoral thesis entitled 'Material development and product development oriented to fused deposition large scale additive manufacturing' in July 2019. He has been a formal lecturer at the Engineering School of Cadiz University since 2015, teaching in the Industrial Design undergraduate degree and focusing his research and projects in the field of additive manufacturing working together in material development and product design. He has undertaken teaching or research stays at The Royal Academy of Arts (UK), Technological University of Munich (Germany), and University of Knoxville in Tennessee (USA).

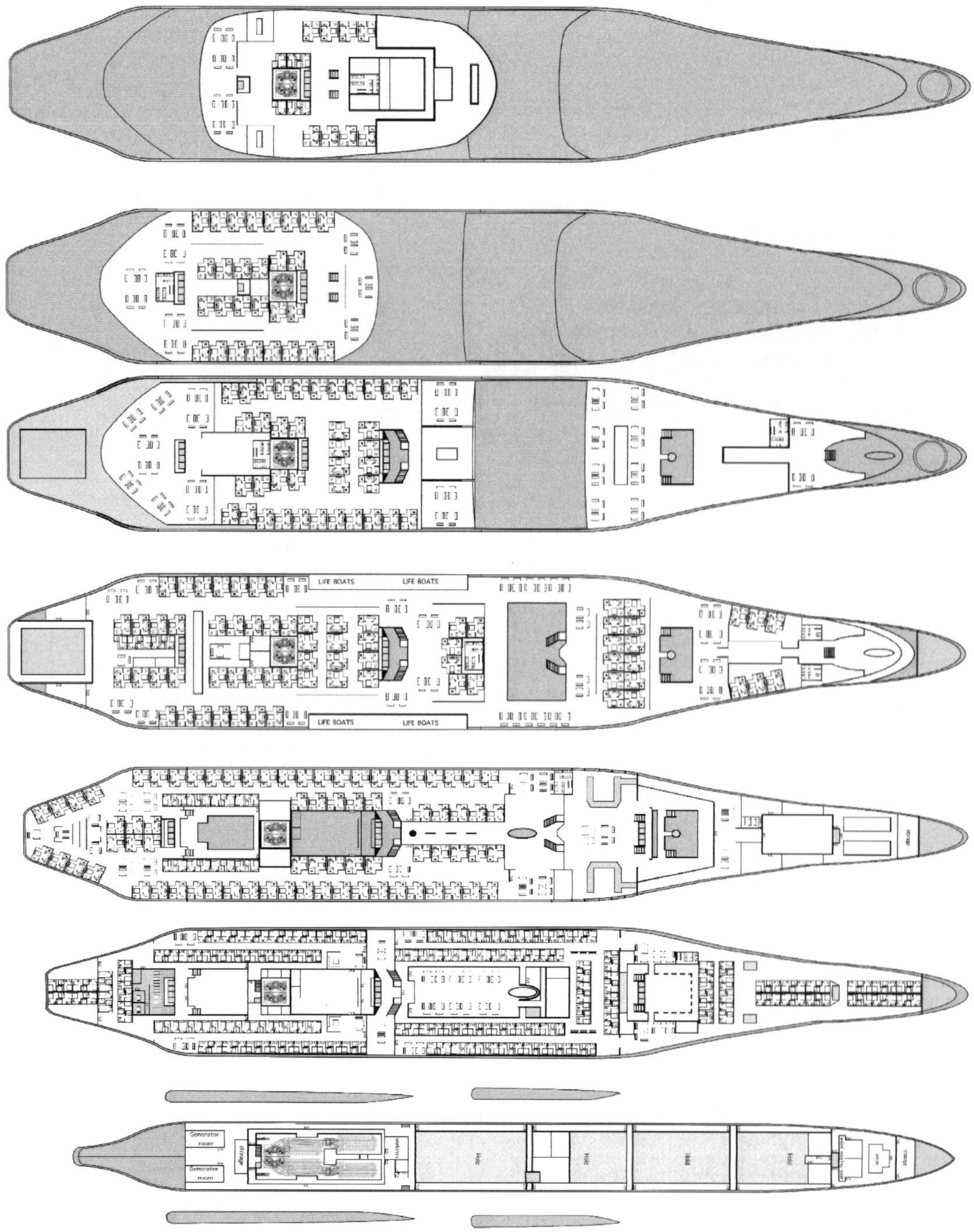

Figure 13: GA of La Trasátlantica

DIGITAL DESIGN IN THE SHIPBUILDING ENVIRONMENT AND THE CLASSIFICATION APPROVAL

R. Pérez, SENER, Ingeniería y Sistemas, Spain

New and innovative AI technologies are developing which will eventually become integrated within digital design systems to facilitate the design process itself. The technologies of Big Data analysis and fast searches will facilitate the design of a part or any vessel concept through the application of design rules. The integration of the design validation process of the 3D models by the Classification Societies will be done by using cloud computing or with direct connection with cloud applications. This will result in faster and more reliable ship design. Ship design also requires the validation of the 3D model from the point of view of the Classification Societies. For this purpose, models have to be exported into a specific format for calculations. There is an additional step that adds complexity to this process: the Classification Societies normally want to review the models by their own means. This paper reviews the latest advances, and CAD Industry efforts, regarding the classification approval process for ship design, through the use of 3D models. This paper discusses the various classification approval process scenarios and describes the solution adopted to be integrated into a CAD System. The digital age has arrived, as an age of collaboration, where new technologies will be imposed and where the enterprises of the shipbuilding industry and the Classification Societies will adapt both their internal processes and their methodology in parallel with the general progress of the industry.

NOMENCLATURE

BV	Bureau Veritas
CAD	Computer Aided Design
DNV	Det Norske Veritas
GL	Germanischer Lloyd
HVAC	Heat, Ventilation and Air Conditioning
IoT	Internet of Things
LR	Lloyd´s Register
ROI	Return On Investment

1. A BIT OF HISTORY

Drawings to define a ship hull became popular only in the seventeenth century in England. It was during the nineteenth century when the Classification Societies work exclusively with the blue prints.

The first Society to print a Register of vessels was Lloyd´s Register (LR) in 1764, to provide an idea of the condition of the ships that they chartered and insured. The LR, as well as Germanischer Lloyd (GL) and others, was named after a seventeenth century coffee house in London owned by Edward Lloyd. Figure 1 shows a recreation of the coffee house.

Figure 1: Recreation of the Edward Lloyd´s coffee shop. Photo taken by the author at the LR Headquarters in 71 Fenchurch Street (London)

Time passed, and the Classification Societies worked continuously to better serve shipyards' and technical offices' data management needs.

New information technologies and digital processes have been implemented in the shipbuilding industry, shipyard 4.0, and, in some areas, are pushing to incorporate new essential features into the Classification Society of the twenty-first century.

The digital transformation of the industry started its new course around the beginning of the 2010s, leading it to specific projects that had materialized in new tools now available to the Shipbuilding Industry. This progress will continue, in parallel with the rest of leading industries [6].

The modifications that the digitalization process implies for the work performed by the Classification Society will affect not only the design but also the fundamental work of the inspections and other digital services that will help and support both shipyards and ship-owners.

2. HOW TO APPLY THE INDUSTRY 4.0 TO THE CAD SHIPBUILDING ENVIRONMENT

The shipbuilding process generates a lot of information and data, which makes it impossible to have all this data in real time, but new processors, simpler and smaller, with a good connection to the Internet, make this possible.

The different machinery equipment to be installed in the ship will have to be prepared for connections to provide the necessary information that the components must to share for normal operation [3].

There should be two different networks of connections. Firstly, an internal network where the systems and components of the ship work together to get the best operation possible. Different smart devices, components and systems will share their state, or their needs for a better operation. But, secondly, another connection must be made with the external world. This external connection will provide the interconnection of the whole ship as a smart ship with the centralized intelligence of the ship-owner. This connection will share information necessary for the better operation of the ship, or will allow the receiving of orders for a better operation according to information received from other ships.

Furthermore, the information necessary to get the replacement or revisions of the components will be transmitted from the components itself.

To make this approach possible it is necessary that the Computer Aided Design (CAD) tools, a symbolic representation of which appears as figure 2, are able to manage the amount of data generated and the relationship between the components and target of the information they share.

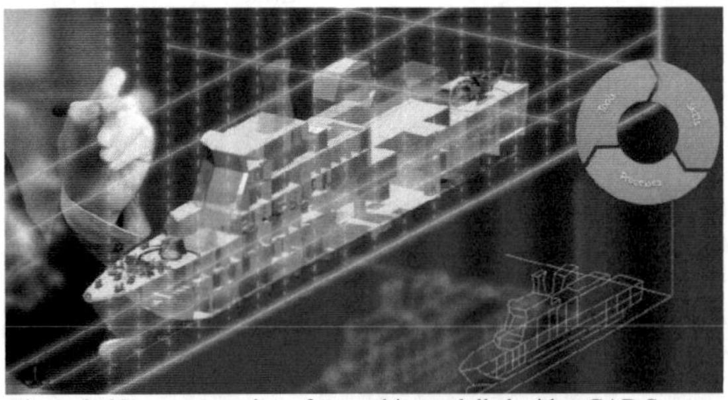

Figure 2: 3D representation of a warship modelled with a CAD System

Adapting shipbuilding to the concept of Internet of Things (IoT) can be done from a variety of approaches [4].

Future ship design tools must be released to take advantage of the information that smart devices and smart ships already adapted to IoT will supply. Information such as the performance of its operations, consumption, trips, etc. They will allow for an early evaluation of naval projects and better designs.

The information shared in the scope of the IoT must be managed alongside the whole lifecycle of the ship, starting from the beginnings of the initial design. This need requires the CAD tools to be prepared with specific characteristics to handle that information.

Many of those characteristics are not yet available and the software developers must programme them. It is necessary to design CAD software able to receive information of ships in operation to evaluate the performance of the designs with real data. And, of course, the product data model must have the ability to incorporate relations between internal and external components.

3. WHAT IS THE DIGITAL TWIN AND HOW MAY IT BE ACHIEVED?

Everybody speaks about the digital transformation but it is not always clearly understood what it means. It may be clarified by describing the three main pillars on which this transformation is built [5]:

- **Digital services**. The digital services include the generation of a Digital Twin of the ship, with all the relevant information. This mock-up is going to be of high value, not only for the Classification Societies, but also for the shipyards during the phase of construction, and for the ship-owners along the exploitation phase.
- **Customer Experience**. Customer experience may be improved by means of digital certification and with a specific set of applications that will facilitate the tasks of inspection and maintenance according to the Classification Society Rules.
- **Operations**. In this case, the enhancement shall focus on the improvement of the effectiveness and efficiency of the interventions on board the ships and in the internal procedures of the Classification Societies.

Cybersecurity (and particularly software piracy), due to technological advances that have enhanced the automatization of ships, and because we are increasingly dependent on communications, together with the new technologies based on software (dynamic positioning, consumption optimization, electronic systems, gas emission reducers, etc.), are making our hardware/software systems potentially one of the weakest points for the international navigation. Therefore they must be enhanced and make secure. Both the Digital Twin and its fundamental benefit, the digital services, have demanded a revaluation of aspects such as cybersecurity, piracy control, accidents prevention, performance of the software or systems, etc.

Inside the concept of cybersecurity, the Classification Societies are defining a new classification notations, such as the SYS-COM from the Bureau Veritas (BV) [5], that is assigned to those ships that will have increased reliability thanks to the communication services between the ships and the land base. This will allow the monitoring analysis as well as the remote resolution of issues thus increasing the reliability of ships and their operations. The use of these automatization systems for support and maintenance is presented in figure 3.

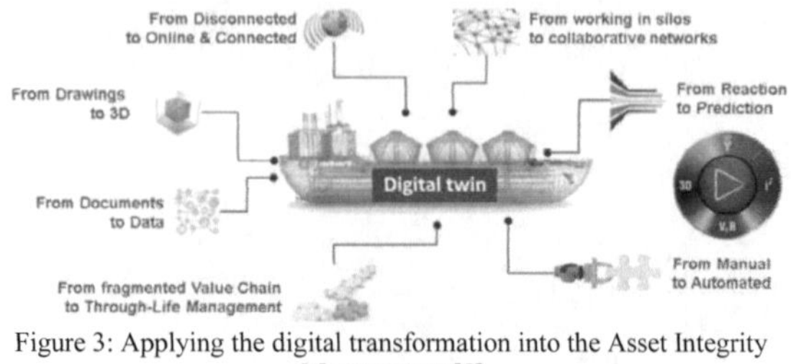

Figure 3: Applying the digital transformation into the Asset Integrity Management [5]

Other tools that are already been in use are those oriented to improving the customer experience of the ship-owner:

- **Mobile applications**. Specific applications developed for ship-owners that can perform the tracking of the ship, handling its certificates, as well as classification services, etc.

- **Specific cargo tools**. Are those where it is possible to verify that a ship is ready to transport a particular cargo. Responding to specific needs of the shipping company to verify, in the short term, the viability of transporting a particular cargo that is not in the originally approved cargo list.
- **E-Commerce**. E-commerce started as a web tool where the ship-owner can request the certification of the Ship drawings, at the same time as obtaining an on-line price for it, having at the same time the possibility of paying with credit card. It is specially designed to improve ship-owners' experience, above all by taking into account the financial advantages of reducing the time to get the Class Certification.

4. THE NEED TO CLASSIFY A 3D MODEL

Recent initiatives include the long being awaited classification of 3D models. This avoids the need to send classification drawings as part the verification process; instead just the electronic 3D model will suffice.

Obviously, there may be some other drawings requested by other regulatory agencies, as well as those that must be available on board, in paper form and available to the captain. These include drawings, book of damage stability, and safety drawings, etc.

The advance inspections methods had been already proved their utility inside the process of the classification (with their actual limitations).

The inspections of the Classification Societies would be decided on board by the inspector by means of its electronic signature and simultaneously verified on a secured certified web.

This will be complemented with the tools that will interact between:

- **Shipyard**. By means of the drawings approval management tool with their comments that will be directly linked to the management systems and with the production outputs.
- **Ship-owners**: It will make the interconnection between the ship-owner systems, especially with the objective of knowing the ship's online status, and those issues relate to the classification certification and the regular inspection visits.

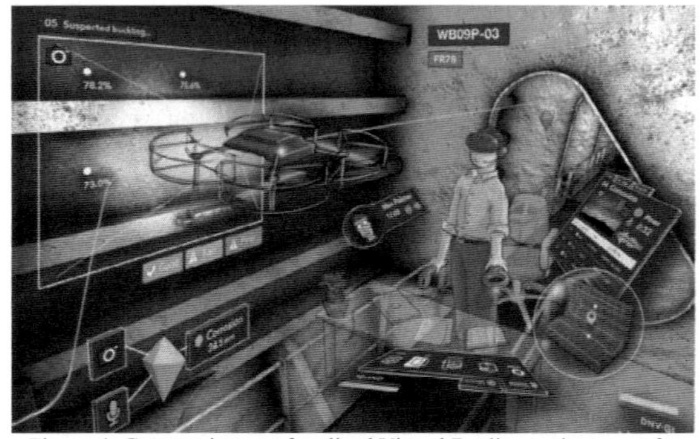

Figure 4: Concept image of realized Virtual Reality environment for remote hull surveys [1]

Currently being finalised are the last functionalities made by the different software firms that will be used by the planners, with the objective of optimizing the available resources of the Classification Societies, giving the best possible service to the ship-owner when the inspection visit has been requested. This request for an inspection visit will be managing by an inspections coordinator that, thanks to the internet planning tools, will allow the appointment of an inspector who would perform the inspection and would issue the corresponding certificates, as is illustrated by the humorous figure 4 from the DNV GL. (For information purposes, GL ceased to exist as an independent entity on September 2013 as a result of its merger with Norway´s Det Norske Veritas (DNV) to become the present-day DNV GL.) Development initiatives will also go different stages, starting with the connecting of the digital model with other CAD systems available on the market, by using a neutral platform that allows the reading of models generated by the different systems and being able to export them back.

A main enhancement to be used by the ship-owners will be the possibility of generating customizable scorecards dashboards with the most significant data demanded by the ship-owners from any particular ship or fleet.

Many other improvement initiatives of the process would surely come from the Classification Societies, and most probably would come from the use of Virtual Reality and Enhanced Reality. Their actual incorporation into the daily base work in the sector is not being questioned as much as *when* they would be available. These topics will be examined in the next section.

5. VIRTUAL REALITY AND HOW TO USE IT

The shipbuilding industry is affected by a global and extremely competitive environment. All processes and systems have to be adapted to challenging scenarios, making special efforts in innovation and applying the most advanced technology available in the market. The application of Virtual Reality to shipbuilding is not new, but it is more extended now thanks to important improvements in software and hardware. From a user point of view, it is possible now to find a wide range of solutions to meet the most demanding requirements, to achieve measurable results, in terms of efficiency and costs.

The 3D ship model is key to all tasks related to ship design and manufacturing. As an example, figure 5 shows how the architecture of Ship Based Virtual Reality Simulator System is possible where 3D modelling is the beginning.

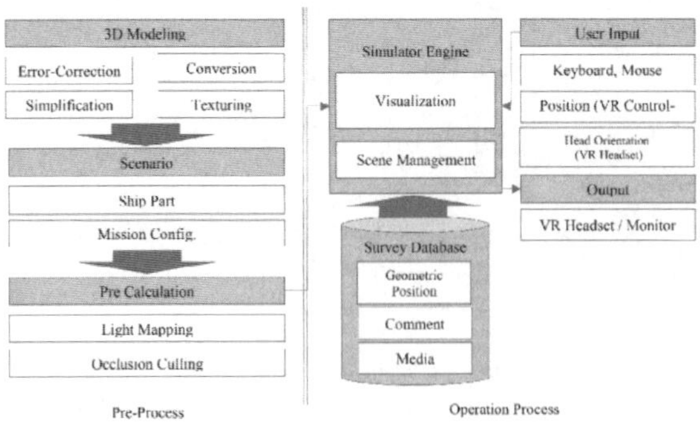

Figure 5: Architecture of Ship Based Virtual Reality Simulator System [2]

It is the starting point of a series of thousands of tasks that are absolutely based on the coherence of the model. As an added value, CAD suppliers develop smart tools to use this model in realistic navigations in Virtual Reality environments with multiple purposes. Some of the most important uses of Virtual Reality in shipbuilding are described in the following paragraphs.

4.1 VIRTUAL REALITY DURING THE SHIP DESIGN STAGES

Maybe the engineering department of a shipyard is where is more extended use of this kind of solutions may be made. The production department is maybe where it should be extended because this is where the most costly errors occur. They don't usually work with Virtual Reality solutions but now it is becoming clear that it is very useful to check the model, not only to avoid errors and inconsistencies but also to improve production tasks.

So, the first and most important use of Virtual Reality is for avoiding errors. The possibility of viewing the 3D ship model, as realistically as possible, is a really effective way to discover errors in all stages of the design. From early design stages, to manufacturing and production phases, it is possible to use it to check all the elements, to see inconsistencies, to prevent interferences and collisions, to question properties and attributes, and to study different design alternatives and changes dynamically. In addition, the user and model interaction has become much more realistic with the implementation of tracking devices which give the sense of being inside the model, walking and moving on it, touching it.

Viewers allow an easy navigation and a fast-movement. Different modes of visualization, search and query options, inclusion of annotations and measurement of distances are important and useful capabilities. Ergonomic aspects in the design can be studied and better checked with Virtual Reality, including the possibility of using dummies. It is also possible to detect collisions, to check interferences, and to insert annotations [7].

Simulation is another important area of application of Virtual Reality in ship design, engineering, and production. It is being applied in the study of many tasks with different purposes, from the study of escape routes to the simulation of dismantling equipment for maintenance purposes. The help in the study of critical assembly tasks and how the surrounding elements are affected is also relevant.

Needless to say, the use of Virtual Reality for the control of the design is most useful, just to see the progress of the project. Moreover, once the ship is being built comparisons between the model and the real ship are necessary and Virtual Reality can help to find out errors that need to be resolved as soon as possible, eliminating further unnecessary costs.

It is very common for ship design and engineering as well as manufacturing to be divided into different blocks that are subcontracted to different design offices, subcontractors, and shipyards. The shipyard receives all the information about the ship and needs to ensure that everything is correct. Having a Virtual Reality tool that allows the integration of all the information in a simple way makes possible a single model, which is much more effective than having many different models [7].

Figure 6 shows an example of a 3D visualization of a complete system using a Virtual Reality advance tool.

Figure 6: 3D representation of a system with FVIEWER®, a Virtual Reality tool

4.2 VIRTUAL REALITY FOR COMMERCIAL USES

The benefits of using Virtual Reality for marketing and commercial purposes are evident. That is why Virtual Reality tools are having a great reception in sales and marketing activities. Thanks to the wide range of possibilities available in the market, from small and portable solutions to big and on-demand solutions, the present and future applications of Virtual Reality are greater than we might currently imagine.

From the marketing point of view, the possibility of presenting a 3D ship model with the higher level of detail in an immersive Virtual Reality experience adds incalculable value to the sales activity. Many shipyards are already taking advantage of this technology, not only in their facilities but also in fairs and exhibitions thanks to the portable solutions that will be described later.

4.3 SHIPYARD MANAGEMENT
Shipyard management plays the role of ensuring that commitments made to clients are fulfilled. The project needs to be accurate, with the highest quality level and in accordance with estimations to ensure customer satisfaction. To achieve this, it is very important to have a clear idea about the progress of the project and also about changes that can affect further requirements, schedules, and costs. Having the opportunity to check the Virtual Reality model from early design stages is a great advantage and allows friendly checking, quick evaluation of alternatives, and fast decision-making. Ultimately, the great impact in cost reduction is possible thanks to early error detection, it being much more expensive to make modifications in the manufacturing and production stages.

4.4 SHIP-OWNERS MANAGEMENT
The presentation of the progress of the ship project to ship-owners is another important use of Virtual Reality, adding the value of making it easy to conceive the project in a very realistic and intuitive way. This has become a great advantage in comparison with conventional presentations. In fact, Virtual Reality can replace the old scale ship models that are expensive and are short on level of detail. The possibilities of presenting the information available in the Virtual Reality solutions are as big as the user´s ability, because now it is possible to provide realistic finishes with all the level of detail required. Ship-owners can also check and monitor the progress of the model and promote design alternatives that are less cost-effective in further stages of the whole process.

4.5 NAVAL SHIPBUILDING MANAGEMENT
Military shipbuilding industry is affected by the most demanding requirements both in the design and production stages. It is true that this kind of naval project, both in submarines and in surface ships, is lengthy and complex although they do have more human resources. But, at the end, the project needs to comply with the strictest rules and also with the budgets imposed by government ministries. This is not an easy task, and that is why all the help in the control and supervision of the process is really welcome. It is in this area of shipbuilding where Virtual Reality solutions have been well received some time ago. The application of the most advanced technologies usually starts in ambitious software tools with major innovations and budgets. That is no longer the case with the use of Virtual Reality in shipbuilding. Even though we cannot forget that having a comprehensive Virtual Reality solution is costly, and perfectly valid for long-term military projects with larger budgets, there are now new technologies such as the Virtual Reality Headset that have dramatically reduced the costs keeping a great performance.

Figure 7: Indication of thickness measurement points and camera positions for close-up imagery viewed in Virtual Reality [4]

Another important use of Virtual Reality in naval shipbuilding is with the navy or merchant fleets. The use of Virtual Reality has great advantages for different reasons, for example, for measuring thickness, as shown in figure 7, or for crew training. In submarines this can be particularly interesting because of lack of space. Simulation is the other important activity, since there it is really difficult to enhance some kind of operations inside a submarine just as much in a surface ship. The help of Virtual Reality to simulate operations and dismantling and maintenance tasks is unquestionable.

4.6 AREAS OF IMPROVEMENT
From another point of view, there are still some drawbacks to mention about the use of this kind of technology in shipbuilding. The most important one is the price. Huge savings can be achieved with Virtual Reality Caves in terms of Return On Investment (ROI) aspects as opposed to just the investment on its own, because although there are alternative devices available in the market, the truth is that Cave hardware is still expensive, and there are additional costs such as software, implementation, room preparation, and the training.

Another important drawback is that it is necessary to have a good 3D model of the ships. Sometimes this is not possible, because there are multiple CAD systems used in the same project, with the difficulty of having just the one model of the ship with all the information. To avoid this situation, it is much better to develop the whole project in the same CAD, or at least to have the necessary tools to integrate the information in a viewer compatible to all of them, which sometimes is a very difficult task.

Finally, the ergonomic aspect of the solution is important too. The use of 3D glasses for a long time is annoying for the eyes, and the standing position looking at a screen without any light could be unpleasant if used too often.

6. 3D MODELS & CLASS SOCIETIES WORKING TOGETHER FOR A BRIGHT FUTURE
Classification Societies and CAD vendors have taken their time preparing oriented solutions including those that allow the minimization of designer errors, as shown in figure 8.

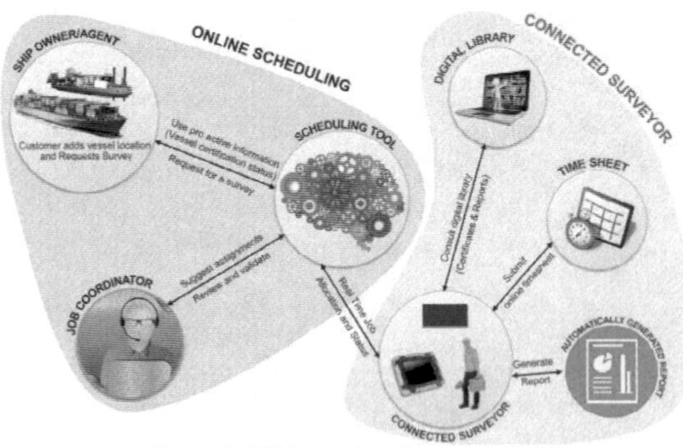

Figure 8: BV inspection planning tools [5]

In recent projects, there has been the need for ship designers to use CAD tools to work better, avoiding the introduction of mistakes into the 3D model, with all parts of information and classification and construction drawings being as accurate and precise as possible. A CAD tool with design rules embedded would enable the reduction of time and cost, and produce detailed 3D models and accurate classification drawings.

During the first stages of ship design is where there are more implications downstream, so more mistakes could imply more loss of productivity. Key to future improvements regarding General Arrangement and Naval Architecture Calculation are:

- The CAD tools should not be enabled to place compartments in incorrect locations, as they are stated on the Rules of the Classification Societies, and they should also request the inclusion of a cofferdam. As an example, the CAD should avoid locating a fuel tank next to the external hull.
- This CAD Systems should avoid the wrong positioning of any item of equipment inside the wrong compartment.
- According to power estimation, the software should apply the most convenient method to calculate the Effective Horse Power, because the programme already has the necessary information about hull forms, for example, Froude Number, Block and Prismatic Coefficients.

In addition, in hull structure discipline there is plenty of room for improvement. Here are some ideas for the future:

- The Systems should avoid the location of macro-holes closer than a minimum distance from a watertight bulkhead.
- The CAD tools should avoid piping, Heat, Ventilation and Air Conditioning (HVAC) and cable trays penetrations if in the structure there is a hole closer than a minimum distance according to the Classification Society's rules.
- Improvements in the welding management. There are so many areas where the designer could make an error applying welding procedures or tasks.

As mentioned before, there are many areas where the CAD Systems could improve the design rules, and the electrical discipline is not an exception:

- The Systems should control the minimum and maximum distance between the cable tray and the surface where the supports lay.
- There is no control of the cable tray material to be used based on the zone of the ship, i.e. steel, aluminium, etc. The CAD tools should control the cable characteristics, as it is isolated against fire, oil, corrosive, low emission in the case of fire.
- There is no control regarding connectors/equipment linked with specifications.

Sensors, Big Data, and the cloud are some of the most interesting changes in the way technology collects and manages data in recent years. These changes have not significantly influenced the common practices in condition monitoring and validating CAD Data. In part this is due to the reduced trust in data security, data ownership issues, lack of technological integration, and the apparent invisibility of direct benefit. It is necessary to design and prepare a method for incorporating smart sensor techniques and distributed processing in data acquisition for condition monitoring to assist during the design process.

New technologies are currently appearing that will very soon be linked with design systems to facilitate the design in self. The technologies of treatment of Big Data and fast searches will allow the engagement of the design of a part or any concept with the relevant rules. The integration of the validation of the structural models by the Classification Societies will be done by using cloud computing or by direct connection with cloud applications. This will result in faster and more reliable design.

Monitored full-scale performance data should be analysed in the future to improve performance at the design and manufacturing stages. Performance evaluations should be made, employing developed monitoring and performance analysis methods, so full-scale performance can be evaluated with a high degree of confidence, and results effectively utilized at the ship design stage. CAD simulation tools and Artificial Intelligence for user guidance, combined with Big Data using Artificial Intelligence to help ship designer for optimizing the 3D model, makes it easy to predict that this is the future.

There are many advantages of using CAD in shipbuilding: ease of design with the design rules embedded, speed of design, and use and reuse of information. It is expected that in the future CAD tools will advance further and allow greater information management through these new improvements. In general, CAD Systems provide tangible benefits while the process is optimized, reducing design time and production, and therefore costs.

7. CONCLUSIONS

As a summary we may say that the digitalization process in the Marine Industry is being adapted to the new demands of the industry 4.0, being focused as much on the internal processes of the shipyards as on the interaction with the rest of partners. Thanks to new technologies, amongst others, the generation of the Digital Twin, has been highlighted. This will permit the monitoring of the life cycle of the ship from the initial design, and through the different stages of the operative life of the ship, right up to its being scrapped.

With reference to the inspections to be made, the Classification Societies are evolving in the two main fields that govern this area:

- Inspection requests and consequently feedback into its Digital Twin.
- The inspection itself, by means of using the new inspection methodology available in our industry (for example, drones).

Notably these inspections will be more concerned with risks (Risk Based Inspections), that will be managed by specific analysis programmes that will evaluate the frequencies (probability of occurrence) and severity of the damages, generating a risk matrix that will allow the elaboration of a planning for the equipment inspections based on these indexes, optimizing the inspection visits and allowing a better planning of them [5].

These steps, aligned with the permanent process of digitization where the Classification Societies and the rest of enterprises of the sector are immersed, and where the final objective is the perfect alignment with the tools and methodologies of the customers, will offer a better and more efficient service in each interaction.

This is an example of how it could be in the short term the process of classification, for a ship-owner [5]:

The ship-owner will be alerted through his mobile phone/computer that the date for the inspection visit will be reached and depending on the warnings that have been configured, the ship-owner will request a visit (in time and location) by means of the same tool, and this information will be transferred in real time to the Classification Societies. The Classification Societies will search the best inspectors available, with the necessary accreditations according with the kind of ship and available in the ship-owner's requested dates and place, and will automatically issue the working order to the designated inspector, all within the mobile phone. Lately the onboard inspection, either conventionally or with the new inspections systems, would generate the reports by means of the same tool, obtaining even the Digital Certification, with the corresponding savings of time in all the different stages of the inspection process.

If it is necessary, the measurements taking on board (for example, thickness), will be stored in the Digital Twin that would automatically perform a recalculation of the ship structure analysis, in case that is necessary, giving the corresponding recommendation, and optimizing in this way certificates renewals. At the same time, the system will integrate all remarks (video, pictures, comments, text, etc.) coming from the inspector, issuing immediately the inspection reports with its corresponding digital signature.

The digital age had arrived, an age of collaboration where the new technologies will be imposed and where the enterprises of the Shipbuilding Industry and the Classification Societies shall adapt both their internal processes and their methodology in parallel with the general progress of the Industry.

8. REFERENCES

[1] CABOS, C.; WOLF, V. and FEINER, P.: *"Remote Hull Surveys with Virtual Reality"*, COMPIT, Cardiff, Wales. pp. 444-453, May, 2017.

[2] KIL, W.S.; SON, M.J. and LEE, J.Y.: *"Development of Multi-Purpose VR Simulator for a Ship from 3D CAD Model"*, COMPIT, Pavone, Italy. pp. 250-258, May, 2018.

[3] MUÑOZ, J.A. and PEREZ, R.: *"CAD tools for designing smart ships in the world of the Internet of Things (IoT)"*, Smart Ship, London, United Kingdom, vol. 1, pp 29-35, January, 2017a.

[4] MUÑOZ, J.A. and PEREZ, R.: *"The Internet of Ships: a new design for Smart Ship"*, The Naval Architect magazine. pp. 26-30, 2017b.

[5] PANCORBO, J.: *"Bureau Veritas: evolution towards a digital industry"*, Revista Ingeniería Naval. pp. 75-80, June, 2017.

[6] PÉREZ, R.: *"The latest advantages in shipbuilding CAD Systems"*, SORTA, Trogir, Croatia. October, 2016.

[7] PEREZ, R. and ALONSO, V.: *"Virtual Reality in a Shipbuilding Environment"*. Advances in Engineering Software. March. Volume 81. pp. 30-40, 2015.

9. AUTHORS' BIOGRAPHIES

Rodrigo Perez holds a PhD in Marine Engineering. He is Head of the Naval Shipbuilding Area at SENER. He is also professor in several subjects at the Marine Engineering School of the Technical University of Madrid.

DESIGN ANALYSIS OF HIGH-SPEED VESSEL

G. L. Krishna, D. Kumar and V. A. Subramanian, Department of Ocean Engineering, IIT Madras, India

SUMMARY

Composite hulls, typically in fibre-glass, are commonly used for high-speed applications. Structural design of the hull requires knowledge of the stresses encountered in the dynamic environment of the sea. It is important to reliably estimate the upper limits of stresses experienced by the hull for safe design as against conservative design with excessive safety factors. Determination of stresses induced by dynamic loads is of prime importance. High-speed planing crafts experience high dynamic loads in the form of pressure peaks and slamming due to encountering waves. This paper presents a methodology to determine dynamic loads on the planing hull form when encountering external waves at high speed. The application of computational fluid dynamics provides the estimation of hydrodynamic loads on the hull form. The approach here uses overset mesh to effectively handle the visualisation of motion and to assess the dynamic effects due to the occurrence of waves and spray. Once the dynamic load distribution is known, an uncoupled fluid-structure interaction approach helps to determine the structural stresses and therefrom the design for safe loads.

NOMENCLATURE

C	Courant number
V	speed of hull in m/s
L	waterline length
Δx	distance of the first cell to the hull surface
Δ	Displacement

1. INTRODUCTION

High-speed crafts are commonly used in a variety of applications such as patrol, sports, leisure, and transportation. An optimum design requires consideration of the aspects of drag, motion characteristics, and dynamic loads which result in structural stresses on the hull form. High-speed vessels are subjected to hydrodynamic slamming pressures in waves. The hull structure suffers impact loads due to slamming, which can also create elevated stresses in the hull. Computational fluid dynamics with proper meshing strategy offers a reliable method to simulate the real flow conditions including generation of spray and vessel dynamic lift and trim and to estimate the hydrodynamic forces on the hull. By simulating different speed conditions in combination with oncoming waves, it is possible to estimate the dynamic pressures experienced at different locations on the hull along its length. Such simulations also provide the main oscillatory motions of heave, roll, and pitch. The responses are a function of the effective wave encounter period. It is also important to examine the effects of slamming since high-speed vessels made of fibre-reinforced material have lower natural frequency which could be nearer to the encounter wave frequency. The loads obtained through the CFD can be transferred to the structural solver and the stresses and deflections developed on the vessel calculated using Finite Element solvers and the effects of slamming pressures can be studied globally and locally. Vessel and waves have relative velocity between them when the vessel moves in waves. When the vessel is at high forward speed in waves slamming is a distinct possibility with attendant high dynamic pressures. The bow region of the hull is likely to experience maximum slamming loads [12]. The present study records pressures at critical locations as a function of the encountering waves.

Earlier studies on high-speed vessels were performed on 2D hulls based on the conservation of momentum by Von Karman [1]. Savitsky [2] developed semi-empirical relations for the calculation of drag and lift. Many researchers studied the high-speed vessels using potential flow theory, which is used to determine the lift and drag coefficients. Most of the empirical methods are not suitable if an unconventional vessel is under consideration. Experimental studies are complex and costly to conduct. Therefore studies based on computational fluid dynamics offer a cost-effective solution. Azcueta [3] conducted a numerical simulation of high-speed crafts using COMET software. Sadik et.al. [4] carried out numerical simulation using Starccm+ to obtain the flow and motion characteristics. k-ω model is used as the turbulence model. Yousefi et.al. [5] reviewed different analytical, experimental, and numerical techniques used for the analysis of high-speed vessels. Lotfi et al. [6] carried out a numerical investigation using CFX on the high-speed stepped planing hull by the Volume of Fluid (VOF) approach. The entire planing range of the vessel is covered and compared; the wetted surface area is analysed with the available experimental data and existing semi-empirical methods. Sukas et al. [7] used the overset grid method in attaining better dynamic equilibrium for the complex fluid-body interaction problem. Sun and Faltisen [8] implemented 2D+t theory for solving the fully non-linear time dependent solution for planing vessel in waves using the boundary element method. Pitch and heave RAO were plotted for different wavelengths and compared with the experiments performed by Fridsma [9].

2. METHODOLOGY

The loads and motions developed on the vessel are obtained using RANSE based CFD solver Starccm+ [10]. It fundamentally solves the continuity and Navier-Stokes Equation which describes the fluid flow over the body. These equations are based on the conservation of mass and momentum respectively. The Reynolds averaged form of the N-S equation (RANSE) gives the solution of governing equations for practical engineering problems. In this method the computation of pressure distributions and velocity are obtained by solving the Navier-Stokes Equation, modelling the free surface and hull position in equilibrium. Free surface is fundamental in defining the flow around the hull. Free surface as the boundary between air and water imposes the dynamic condition, and the kinematic condition fulfils the physics of the flow. The Volume of Fluid (VOF) technique has been employed for the modelling of the free surface. The Dynamic Fluid Body Interaction (DFBI) module effective couples the fluid flow and body motion. It simulates the motion of a rigid body in response to the pressure and shear forces the fluid exerts, which governs the degree of freedom, mass, equilibrium position of the vessel, and the moment of inertia of the system. Besides the forward velocity the vessel is given two degrees of freedom of motion namely, sinkage (heave), and trim (pitch).

Table 1 Main particulars of the candidate vessel

Properties	Unit	Values
Length overall (Lo)	m	21.85
Length moulded	m	20.5
Length of waterline (l)	m	18.36
Beam	m	5.3
Depth	m	3.25
Draft	m	1.062
Displacement, Δ	kg	46000
LCG from aft	m	7.45
Water Plane Area	m^2	12.09
Design speed, V	knots	25
Block coefficient		0.434

3. COMPUTATIONAL DOMAIN

The primary consideration in setting up a numerical model is to identify the extent of the computational domain and boundary conditions. It is necessary to identify the artificial limits of the solution domain and also the associated boundary conditions. A rectangular domain is used here. The domain conforms to those recommended in the literature and also with the recommendations of the International Towing Tank Conference (ITTC) [11]. As per these standards the computation grid extends to 2L (where L is the length of the ship) in front of the ship, 6L behind the transom, and 2L above the hull deck, below the keel, and to the sides of the hull. Due to the centreline symmetry of the ship as well as the flow the modelling is restricted to half hull in the domain. This size was sufficient to get a domain independent solution. In the present study, velocity inlet, pressure outlet, wall with no-slip (over the hull surface), and wall with slip (at the far boundaries) conditions are enforced in the domain. Solver settings and boundary conditions are given in Table.2 and Table.3 respectively.

Table 2 Solver settings

Parameter	Settings
Solver	3D segregated flow, Implicit unsteady
Turbulent model	Realizable k-ε
Multiphase flow model	VOF
Motion solver	Dynamic fluid body interaction (DFBI)
Pressure velocity coupling	Simple
Wall treatment	Two-layer all wall y+ treatment

Table 3 Boundary conditions

Surface	Boundary condition
Inlet, Bottom, Top, Side	Velocity inlet
Outlet	Pressure outlet
Symmetry	Symmetry plane
Hull	body Wall (no-slip)

Time-step

The time step should be small enough to capture the physics of the problem. Time step is selected based on the CFL (Courant- Friedrich-Levy) condition.

$$C = \frac{V * \Delta t}{\Delta x}$$

Courant number should be less than 0.7 for first order time discretization. As per ITTC guidelines [11], the time step should be taken based on an equation, which is a function of waterline length and hull speed.

$$\Delta t = 0.01 - 0.005 - \frac{l}{V}$$

Overset Mesh

Overset meshes, also known as Chimera or overlapping meshes, are preferred when the body is moving with several overlapping meshes or multiple moving bodies. In an overset mesh, cells are grouped into active, inactive, or acceptor cells. Within active cells, discretised governing equations are solved. An overset mesh has two regions, a background region (static), and the overset region (moving). An overset mesh is useful for moving bodies: the meshes near the body are neither morphed nor remeshed to capture the motion. The mesh generation was suitably refined near the hull region and free surface areas to capture the motions and free surface boundary. The linear interpolation option is employed for interfacing the overset and background regions.

The hydrodynamic analysis of the high-speed vessel is carried out in waves with different wavelengths at the design speed of the vessel. Since the waves and ship are moving the effective wavelength of the wave decreases. During the analysis only pitch and heave motions are allowed; all other motions were arrested. Motion responses in different wavelengths are depicted in Table 4.

Table 4 Motion response of vessel in waves

sl No	Wavelength	Waveheight	Pitch (deg)	Pitch (Rad)	Heave (m)	Heave RAO	Pitch RAO
1	4 L	1.6	4.28	0.07	0.95	1.18	1.188
2	3.5 L	1.4	4.27	0.07	0.85	1.22	1.185
3	3 L	1.2	4.23	0.07	0.72	1.19	1.175
4	2.5 L	1	3.91	0.07	0.56	1.12	1.086
5	2.25 L	0.9	3.52	0.06	0.47	1.04	0.977
6	2 L	0.8	2.91	0.05	0.34	0.86	0.808
7	1.75 L	0.7	2.19	0.04	0.23	0.67	0.608
8	1.5 L	0.6	1.45	0.03	0.14	0.46	0.403
9	1 L	0.4	0.55	0.01	0.04	0.22	0.152

Table 5 Dynamic pressure along the length of the hull

Sl. No	POINTS	4 L	3.5 L	3 L	2.5 L	2 L	1.5 L	1 L
		kPa	kPa	kPa	kPa	kPa	kPa	kPa
1	1	14.7	13.6	14.0	15.0	14.1	12.7	11.4
2	2	16.6	14.1	14.6	15.5	14.2	13.0	11.5
3	3	16.4	13.3	13.9	15.0	13.8	12.8	11.5
4	4	15.3	12.6	13.1	14.3	13.2	12.5	11.2
5	5	13.7	12.2	12.6	18.3	12.8	12.3	11.1
6	6	12.1	13.2	14.5	24.2	17.0	12.8	10.4
7	7	13.9	15.7	16.9	28.0	22.2	15.2	10.3
8	7.5	18.8	20.7	22.7	34.7	28.9	22.2	14.5
9	8	30.5	32.4	35.7	50.0	60.3	40.5	28.6
10	8.5	37.6	40.9	45.3	61.1	44.0	54.6	45.7
11	9	39.7	43.1	47.5	61.5	64.7	59.0	0.2
12	9.5	0.7	1.2	0.7	63.6	39.5	0.2	0.2
13	10	0.7	1.2	0.7	0.2	0.2	0.2	0.2

Maximum pitch and heave RAO (Response Amplitude Operator) are obtained for 4L and 3.5 L waves respectively. The values of pitch and heave obtained through CFD are compared with the values of experimental and numerical results conducted by Fridsma [9] and Sun and Faltinsen [8] respectively. The hull used for comparison has the same waterline length to breadth ratio and deadrise angle. The trend of the heave and pitch RAO are matching, and values are also within the range. Pitch and heave RAOs are depicted in Fig.1 and Fig2. respectively.

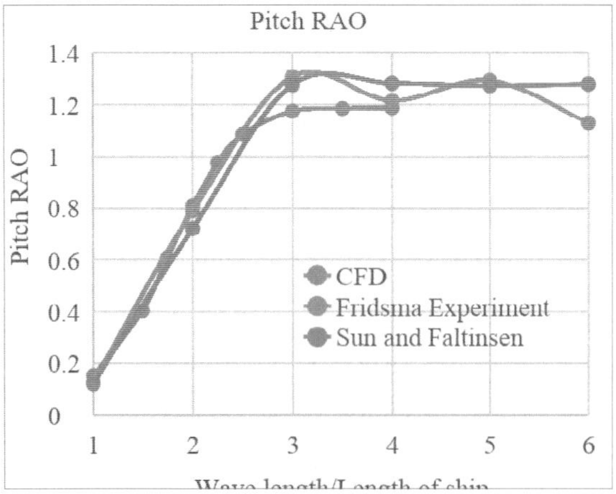

Figure. 1 Pitch response

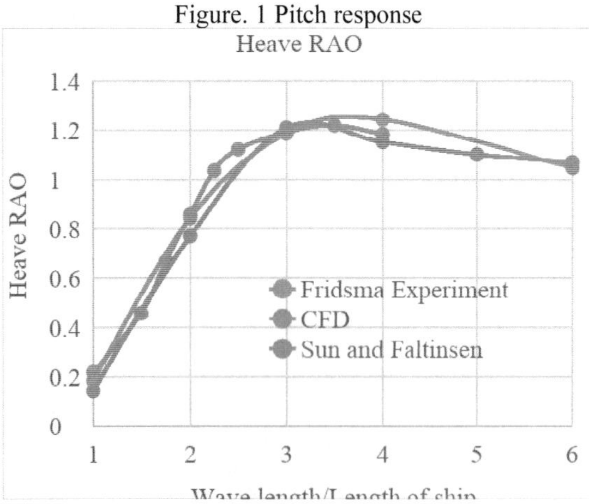

Figure. 2 Heave response

Dynamic pressure developed on the hull when it encounters different waves is recorded at various points along the length of the hull. Pressure probes are created along the keel every 2m; in the forward region it is every 1m. Pressures at the various points along the hull are depicted in Table.5

4. CONCLUSION

The responses of high-speed vessels in head sea condition are obtained for a range of head sea wave lengths. The results are compared with numerical and experimental schemes available in the literature. The results validate the approach and therefore this tool provides reliable results for further analysis as input structural load in the hull and to obtain the dynamic stresses. The results show the occurrence of maximum pressures at specific regions and the most critical wavelength at which they occur. In this specific case, the maximum pressure occurs for wavelength between 2 and 2.5 times the length of the hull.

5. REFERENCES

1. von Karman, T.,"The impact of seaplane floats during landing," N.A.C.A. TN321, Washington, 1929.
2. Savitsky.D, The Hydrodynamic design of Planing Hulls. *Marine Technology,* 1(1):71-95 (1964).
3. Azcueta, R. (2000) 'Ship resistance prediction by free-surface RANS computations', *Ship Technology Research*, 47, no. 2, pp. 47-62.
4. ÖZÜM S, ŞENER B and ÜNLÜGENÇOĞLU K (2010) 'Resistance prediction of high speed craft using CFD', *Industrial Maritime Engineering.*
5. R. Yousefi, R. Shafaghat, M. Shakeri,'Hydrodynamic analysis techniques for high-speed planing hulls', *Appl. Ocean Res.*, 42 (August) (2013), pp. 105-113
6. Lotfi, P., Kowsari, E. and Ashrafizzadeh (2015) 'Numerical investigation of stepped planing hull in calm water', *Ocean Engineering*, 94, pp. 103-110.
7. Omer F S, Omer K, K., Ferdi, C. and Metin Kemal, G. (2017),'Hydrodynamic assessment of planing hulls using overset grids', *Applied Ocean Research*, 65, pp. 35-46.
8. H. Sun, O.M. Faltinsen, 'Numerical study of planing vessels in waves', *J. Hydrodyn.*, 22 (1) (2010), pp. 451-458
9. Fridsma G. 'A systematic study of the rough-water performance of planing boats[C]'. *Davidson Laboratory Report R-1275*, 1969.
10. Cd-Adapco (2014) STAR-CCM+ v12.06 Manual.
11. ITTC Specialist Committee on CFD in Marine Hydrodynamics, "Specialist Committee on CFD in Marine Hydrodynamics Final report and Recommendations to the 27 th ITTC," 2014.
12. Faltinsen.O.M, Hydrodynamics of High Speed Marine Vehicles, *Cambridge University Press,*(2005).

AUTHORS' BIOGRAPHY

Lal Krishna G is a PhD Research scholar in the Department of Ocean Engineering, IIT Madras, working in the areas of high-speed vessels, composites, fluid structure interaction, and low velocity impact.

Dr. Deepak Kumar is Associate Professor in the Department of Ocean Engineering, IIT Madras. He works in the area of structural dynamics of offshore structures.

Prof. V Anantha Subramanian is a consultant Naval Architect and works in the design, optimization, propulsion, and powering of marine vehicles. He used to work in the Department of Ocean Engineering, IIT Madras.

EDUCATING FUTURE SUPERYACHT DESIGNERS AT SOLENT UNIVERSITY – MSC SUPERYACHT DESIGN

G. S. Barkley, Solent University, UK

SUMMARY

Building on the success of the BEng (Hons) Yacht & Powercraft Design degree and the BEng (Hons) Yacht Design & Production degree, Solent University recently launched the MSc Superyacht Design postgraduate degree. The course adds to student's existing knowledge gained from a wide range of technical backgrounds the design skills and practice associated with Superyachts 24-85 metres long. Students develop skills in Naval Architecture, Structural Design, Design Principles, Design Analysis, and Manufacturing Management through experiential learning during the first two semesters. They solve design problems ranging from space planning using MLO requirements through to FEA and CFD analysis of specific areas. Students work alone and in groups replicating industry practice. For the third semester, students complete an industry placement with a Superyacht Design company or build facility where they develop a research-based project based on a requirement of the sponsoring company. This paper gives an overview of the program and outlines examples of the experiential learning gained by the first cohort.

1. INTRODUCTION

1.1 YACHT DESIGN AT SOLENT UNIVERSITY

This year, Solent University celebrated its fiftieth anniversary of delivering Yacht Design and Manufacturing courses to students wishing to enter the small craft industry. Over the years these courses have equipped graduates, many of whom have become some of the worlds' leading designers and production managers, with the skills required to develop and build many of the world's small craft. Types of craft include luxury sailing and motor yachts as well as small commercial craft including pilot vessels, workboats, and more recently wind farm support vessels.

On reviewing the size of leisure craft over the last fifty years, it is easy to see that whilst back in the 1970s a large 'pleasure craft' was somewhere between 10 to 15 metres in length, today this is considered nothing special. In the 1990s a craft of 18-20 metres was considered a luxury. Today whilst a few designers try to restrict craft length to less than 24 metres to meet the less onerous requirements of the Recreational Craft Directive together with the relevant structural and stability ISO standards, many designers and companies are building pleasure craft that lie between 30 metres and up to 85 metres in length. These craft are generally referred to as either Large Yachts or Superyachts.

These larger yachts over 24 metres are a lot more complex in layout, naval architectural requirements, structural design, system design, and compartmentation requirements. All these vessels need to comply with many statutory requirements such as those laid out by originally, the Large Yacht 1, 2 and 3 codes and more recently the Red Ensign Group's REG Yacht Codes Part A or B.

Many of today's professionals, now 'key players' involved in the Superyacht design and build sector have at some time completed an Higher National Diploma or undergraduate degree at Solent University. These alumni have had the luxury of 'growing up' with the increasing size of these yachts, crossing seamlessly from sub 24 metres to above 24 metres navigating rule changes as they have appeared. The industry now needs to employ graduates, ideally prepared with relevant skills and background, directly into this sector.

1.2 RATIONALE FOR THE MASTER'S PROGRAM

Recent trends show that some of our Yacht & Powercraft Design and Yacht Design & Production graduates are starting careers in the Superyacht design and production sector of the industry. Currently these graduates are what would be termed as 'industry prepared' to work in the sub 24 metre sector of the market. Therefore, although some graduates have gained employment directly in the industry, they are presented with a steep learning curve due to the complexity of craft above 24 metres in length. Equally, employers are losing valuable time training graduates. This is not an efficient use of time for either the graduate or employer.

The yacht engineering course team are occasionally approached by graduates from other technical or engineering backgrounds who want to gain a qualification in yacht design which is recognised by a professional body such as RINA so they could also work within this sector.

In 2016 Solent University began developing the Master's Program titled the MSc Superyacht Design which would better prepare Solent University's own graduates as well as others with the skills needed to kick-start a career in the Superyacht design sector. Solent University received its first intake of students onto the validated course in 2018.

2. THE COURSE ETHOS

2.1 BRINGING TOGETHER SPECIALISMS

Designers and Technologists employed in the over-arching employment of Superyacht design & build have a range of skillsets. The success of a Superyacht project often depends upon how successfully a team involved in one specific area of design and manufacture has understood and interpreted the restrictive limitations another team working on a different aspect of the project has to work within.

An example of this could be that the team developing the General Arrangement and interior design will interact alongside the structural design team ensuring that the deck girders do not lead to an intrusive structure possibly requiring pillars just where the interior design team do not want them. The same could be said for system design.

Summarising, whilst each team has their own speciality, it is hugely beneficial for the design company and its individual team members to have an appreciation, understanding, and experience of the specialisms required to develop the amazing Superyachts we see today.

Thus, the over-arching aim of the Master's degree is to ensure that graduates arriving with different qualifications and skillsets get an opportunity to practise the range of skills required for successful Superyacht design, ranging from the initial concept design through to the final production detail

Students eligible to study the master's degree will typically be drawn from a range of subject specialisms including Naval Architecture, Mechanical Engineering, Structural Engineering, and in some instances from technology-based Architecture programs. Equivalent qualifications will also be accepted from other streams such as superyacht crew wishing to return ashore, providing they have seafaring Master's or Chief Engineer's tickets and time and rank at sea.

3. DEVELOPING THE MASTER'S DEGREE

3.1 COURSE CRITERIA

3.1a Meeting University Requirements

The aim of the Master's program is to ensure graduates feel confident, enabled, and embodied with an appropriate skillset to enter the Superyacht Design industry. To that end, the course must deliver a modern, relevant, skillset, ensuring employability and enterprise are embodied within its heart. Graduates of the course must be equipped with highly transferable engineering design, problem solving and 'soft' skills, as expected by the requirements of the university, potential employers, and RINA.

When developing the Superyacht Design course, it was imperative that it met the requirements imposed by University Regulations, which are informed by the Council for Engineering, and by the Quality Assurance Agency (QAA). These requirements ensure students achieve specific standards commensurate to the level expected of a Master's course. The requirements, stated in terms of learning outcomes which graduates must achieve, ensure that they are equipped with a suitable theoretical, practical and industry-based skillset expected on entering the relevant profession.

3.1b Meeting Industry Expectations

To ensure the curriculum also meets the requirements of industry, the course development process included a dialogue between lecturers and key industry stakeholders including both designers and production managers in the Superyacht industry.

This ensured that current subject knowledge and design practice are embedded in the DNA of the program. It also ensures that relevant engineering software, statutory design codes, and manufacturing technologies are incorporated.

These dialogues were ongoing throughout the development of the course. The results of these discussions were embedded in the course syllabus.

3.1c Professional Recognition

Equally important is ensuring that the course will obtain accreditation from a relevant professional body. For this course, it is likely to be the Royal Institution of Naval Architects (RINA). This will ensure that both the course and more importantly its graduates gain professional recognition and hence improve their chances of obtaining employment in this sector.

Bearing the above in mind, the course and each unit was developed to deliver the theoretical knowledge required of a Superyacht designer. The theory is reinforced in each unit by using practical design case studies and tasks typical to those met in Superyacht industry practice.

3.2 DEVELOPING STUDENT SKILLS

The underpinning remit of this degree is to give students appropriate design experience so they hit the ground running once employed. This relies on course participants gaining 'authentic learning experiences' typical of the type of design and production tasks they will encounter once in commercial practice. Within each unit syllabus both delivery and assessment methods are developed to ensure alignment with modern pedagogic practice. This includes ensuring that assignments set in each unit engage and maximize the potential for students to develop both the scholarly knowledge and practical design skills associated with industry practice.

A wide range of project work and suitable assessment strategies appropriate to each topic ensures that completed coursework achieves the level required by industry. Students produce a mixture of case studies, reports, and portfolio-based work which sometimes involves group work. These portfolios tend to include design drawings, technical summaries and, on occasion, presentations. The work may be based on computer program output ranging from stability and seakeeping analysis through to 3D modelling outputs. This helps develop familiarity with some of the commercial software students may use in employment and also promotes the wide skill base expected of graduates in industry.

With this course the pedagogic skills used in teaching and learning respects the diversity of student learning methods and promotes equal opportunity for all learners.

4. THE COURSE SYLLABUS

4.1 AN OVERVIEW

The full-time course is spread over twelve months, and split into three distinctive 4-month semesters. In each of the first two semesters students follow three subject specialisms per semester. The students are expected to gain an industry placement for the final semester. In this period they work alongside a company on a research or design development project for which they write up a technical thesis. If the work completed is sensitive to an individual company, the thesis remains confidential between the participating master's student, the company, the academic lead, and assessors.

The following sections outline each syllabus and shows some work completed by the students.

4.2 SEMESTER ONE

4.2a Naval Architecture

Students study Naval Architecture with a particular emphasis on hull parametric design and stability. Students are expected to apply this knowledge to Superyacht Design through using the REG Yacht Code Section 11 stability regulations. Students also learn to assess hull resistance using tank testing methods or an appropriate Systematic Series. Students also develop the skills required to optimise propeller design and calculate the required engine power and fuel capacity to ensure speed and range requirements are met.

As some Superyachts are sailing yachts, students also gain an overview of the Aerodynamics and Hydrodynamics of sailing together with an understanding of sailing yacht Velocity Prediction Programs.

Accompanying the theoretical knowledge is a set of weekly classes to get all students up to speed on a range of commercially available industry hull development and analysis software including lines fairing, stability, seakeeping powering, and VPP software.

The unit assessment is based on a portfolio of design exercises which they undertake throughout the unit. Students who have studied Naval Architecture before are encouraged to extend their knowledge beyond that of their understanding before and report this in their portfolio. Students new to Naval Architecture have a steep learning curve to cover the topics outlined above and gain a lot through writing up their portfolio.

The knowledge gained in this unit is utilised in other units such as Superyacht Design Principles and Superyacht Design Realisation, as would be the expected case in industry.

4.2b Superyacht Technology

This unit is split evenly between understanding and developing both the structural design and system design of the vessel.

Students will initially develop a knowledge and understanding of both the global loading and local loadings the vessel has to withstand. Once understood, they will be encouraged to develop Steel, Alloy & FRP Structural Arrangements for various superyachts utilising hull scantling rules provided by classification societies such as the Special Service Craft and scantling rules provided by Lloyd's Register. An example of a structural midship section produced as part of a group project using LR SSC rules is provided in Figure 1.

The second half of this unit concentrates on developing an understanding of the large number of Engine Room systems. These range from diesel to hybrid-power generating systems including their transmission system. Other systems on board modern superyachts are studied and include HVAC, hydraulic, and electrical, piping, and stabiliser systems.

Students will gain industry practice of system design by developing various systems appreciating the need to successfully align and integrate the proposed system within the vessel's structural design and GA.

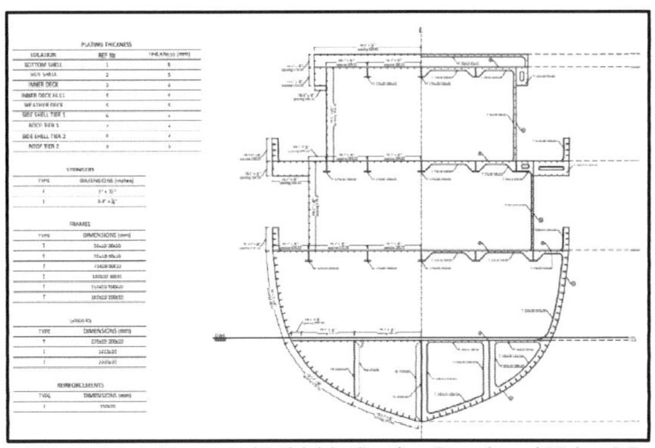

Figure 1.0 A Structural Midship Section Developed Using LR SSC Rules.

4.2c Superyacht Design Principles

Students gain experience and practice of design and utilise many of the industry standard superyacht design CAD packages. Success in this unit is vital as it provides the 'real world' learning opportunities and allows students to develop those skills expected in industry. As a result, this unit in terms of grade weighting is equivalent to the two described units above.

The design tools used in this unit include 2D drafting and 3D parametric modelling software, and also the specific naval architecture-based software taught in the Naval Architecture unit.

The drafting programs will be used to develop new designs up to 500 GT. Students will use the Naval Architecture software programs to assess whether the vessel meets stability requirements as per the REG Yacht Code, powering requirements, sailing yacht performance, or motion sickness indicators. Students will also need to ensure MLO crew space limitations are met with reference to professional crew spaces. Other general design factors include assessing requirements for and implementing fire proofing.

For this unit each student develops a design based on a set client specification. They develop a parametric model and then either develop or modify a lines plan which matches their parametric data. Subsequently, they develop and fit a general arrangement based on the client's specification ensuring the preliminary design satisfies specified REG Yacht code statutory requirements, and in some instances, specific classification requirements.

To summarise: students gain the skills required for industry practice in this unit. As an example, one assignment tasks the student to generate a lines plan, a space plan, and a general arrangement for a sailing yacht between 40-45 meters. The yacht carries four crew and eight guests. Each crew member must be allocated a cabin which is sized to meet the MLO requirements. The client's specification asks for a GA so the crew areas have to be separate from the guest areas. The vessel must also satisfy intact and damage stability criteria. This work is completed over the first eight weeks of starting the course. A sample of work showing the general arrangement in both plan and profile view completed can be seen in Figure 2 below.

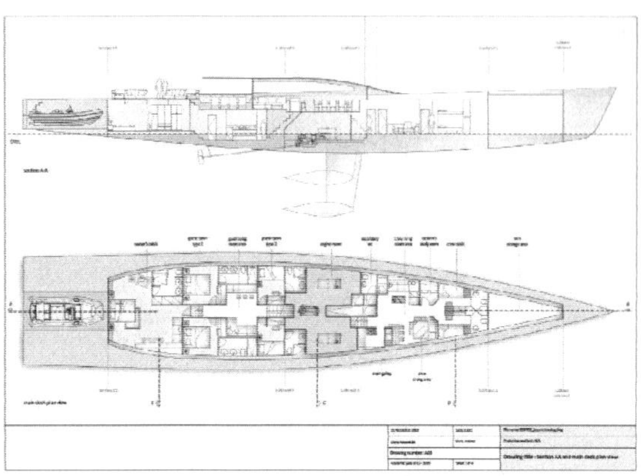

Figure 2.0 Sample of Student Work.

4.2 SEMESTER TWO

4.2a Superyacht Design Analysis

This unit introduces students to the higher end tools associated with Superyacht design.

On the structural design side students learn how to calculate both local and global loads imposed on superyachts for both isotropic and anisotropic materials such as a composite carbon structure.

These loads are then taken a step further using Finite Element Analysis which is often applied to specific areas of a superyachts' structure where direct calculations or those imposed by the ship classification societies are not detailed enough. Students learn the basics of FEA. These are then applied to an area of the yacht, such as a large opening in the hull or transom, to evaluate the forces and stresses in components when subjected to a specific load case. Students learn to validate the solution, comparing the FEA results to those of a simple hand calculation.

Superyacht design often benefits from a knowledge of the likely fluid flow and hence drag or flow speed of either water over an underwater appendage or wind moving over and around a superstructure. Students learn to use a commercial CFD package, gaining experience of developing practical meshing techniques so as to accurately calculate pressures and flow velocities around a shape. This short course covers the practical limitations of CFD and emphasises the importance of validating the computed result against more traditional analytical methods for fluid flow such as boundary layer theory, or where possible traditional tank test results.

From a practical analytical perspective, students get a chance to test either a sailing yacht or motor boat model in Solent University's 60 metre tow tank. Students scale up hull resistance and keel side forces from model to full size to establish the required power in the case of motor yachts or for sailing yachts, heel, and induced drag coefficients to overwrite into a VPP. Students can also test a craft in waves and consider the seakeeping performance.

To summarise: in this unit students gain the theoretical knowledge and practical experience of using these high-end design tools. Most importantly they learn the foibles associated with each analytical method. Assignments for this unit involve writing technical reports, often involving the validation of data from program outputs. This will be a useful attribute to have when interpreting the results of technical reports in years to come.

4.2b Superyacht Project Management

The success of a Superyacht's design and build cycle is normally related to the management of the process. Classical management theory is taught starting with organisational and operational structures, lean manufacturing process, project scheduling, critical path analysis, progress monitoring, and procurement planning.

This classical theory is underpinned by relating this to the project management of superyacht design, including controlling detailed design development and finally to the management of the build process. To this end, students study shipyard layouts, including boat moving and launching arrangements, to name but a few areas.

This unit is delivered initially through a series of traditional management lectures and seminars. These are in turn related to the Superyacht industry utilising case study material developed and delivered by our industry experienced staff who have been instrumental over many years in setting up new Superyacht Shipyards and have also project managed the build of large vessels including superyachts.

To add to the strength of this unit, guest lecturers are invited into the university to give seminars on specific subject areas such as marine contract law. Students also gain from industry visits to local large yacht builders to see theory working as practice. Students gain an understanding of the scale of these projects and the importance of managing the process correctly.

To summarise: students will then apply a part of the project management process for their coursework, allowing them to gain as much of a 'hands-on' approach to the topic of Project Management' as is practicable within a university environment. As a result of this unit, students armed with this knowledge have completed successful project placements in shipyards concentrating purely on this subject area.

4.2c Superyacht Design Realisation

The syllabus builds on the knowledge learned earlier in the Superyacht Design Principles unit, providing students with further skills and tools required to develop the larger and often more complex superyachts. Students go deeper into the REG Yacht Code regulations than before and, on the design side, acquire the skills necessary to drive both technical design and 3D modelling software such as Seakeeping software and Solidworks.

The over-arching theme of this unit is to simulate as close as possible within an educational environment a genuine design office environment experience. This is achieved by the students working in groups to develop a new design which meets the needs of a client and their technical team. By working as a multidisciplinary group of students, each group member can excel in a specific area of design whilst relating their design aspect to the whole project and also gain and practise the soft skills involved with teamwork. The assignment set is split into two distinct phases.

For the first stage, the students were given a client brief for which they had six weeks to develop the original design to a level suitable for a client presentation. The work presented to the client at this stage was very much biased towards the client's brief in terms of a client's typical understanding of 'design' which includes the exterior styling, hotel spaces, and interior styling, and providing space for the client's 'toys'. Prior to presenting this work, the team will have to ensure the vessel is likely to pass certain criteria such as stability and range criteria. The students were graded based on the design presented to the client and the standard of the presentation and also the technical background checks.

The second stage of the project was to develop a technical proposal up to a level expected for a tender document which the designer could present to a bidding ship builder. This document included a detailed weights and centres estimate, ensured statutory stability requirements were met, proposed suitable hull, deck, and superstructure scantlings, and provided details of main propulsive systems for the vessel, to name but a few. This work was written up as a tender document (with rationale for particular decisions) and submitted together with a set of technical CAD drawings. On completion of this work, students met with the 'shipyard team' to discuss the submitted documentation. On the academic front, this provided an excellent opportunity for feedback, and also allowed for a fairer assessment of the student intentions within their proposal.

An example of the first group project was the design of a 70 metre Sub 1000 GT Motor Superyacht for a young rock star client with very specific tastes.

Initially the group had to produce a suitable design and also include exterior and interior renders for the vessel. These are shown in Figure 3 and 4 below.

Figure 3 – Exterior Render of 70m Superyacht

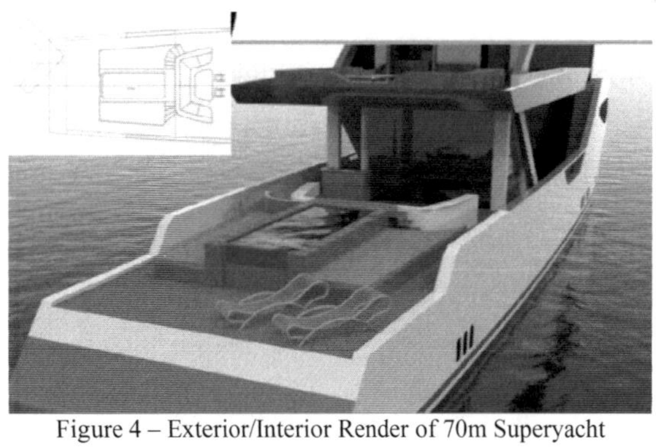

Figure 4 – Exterior/Interior Render of 70m Superyacht

The team also had to develop the interior layout as well as the weights and centres and the tank plan. On this basis a stability analysis could then be completed. Figure 5 shows tank locations for the vessel.

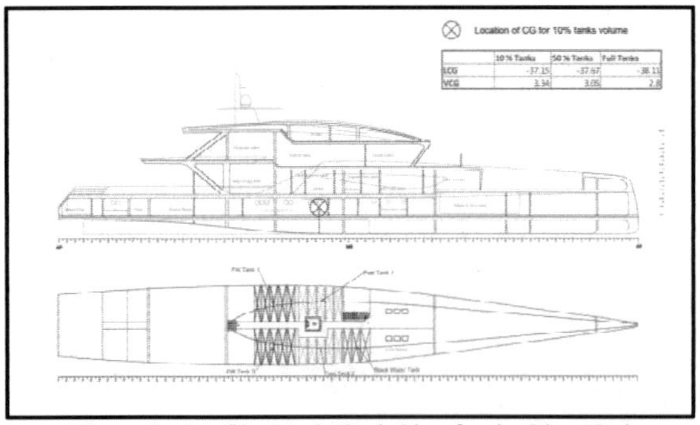

Figure 5 – Profile GA & Tank Plan for the 70m Yacht

One specific area the client was keen on was to reduce the carbon footprint of the vessel. One group member specifically investigated various hybrid solutions for the vessel. The group member investigated various systems, comparing the size of engine rooms, weights of the engines and generators, as well as fuel loads and batteries, and also calculating the overall efficiency of the different drive trains. Figure 6 shows a schematic diagram of a combination diesel electric proposal and also includes a summary of the efficiency of the drivetrain including the propeller efficiency.

Summarising, this unit and the associated assignment provided an excellent opportunity for the student designers to gain some experience of the designer/client and designer/shipyard interface. It also gave students an appreciation of the level of design expected and also the complexity of design and technical design experience they need to work towards once working in industry.

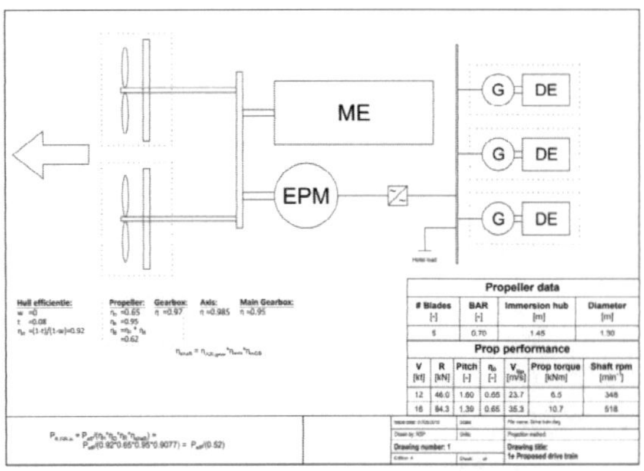

Figure 6 – Diesel Electric Drivetrain Efficiencies.

The project also gave students an opportunity to experience working as a team and occasionally face challenging group dynamics. They also had to work with an unknown client to develop the design, and used the lecturing staff as the technical representative and shipyard interface to add a little reality to the project. An expert from the MCA also made himself available to answer questions regarding the statutory REG Yacht Code.

Making design decisions as a team proved quite challenging. On reflection, the students thought the project proved to be tougher than originally envisaged and that group dynamics could have been better handled. In reality, developing these soft skills has helped prepare them for real world industry.

4.3 SEMESTER THREE

4.3a Industry Based Project

The students spend the third semester completing a master's level project in conjunction with an industry partner. This can be either on site with them, or split between the university and partner. The value of involving an industry partner is that students gain real problem-solving skills based on real industry's needs. Equally the industry gains the knowledge of the work undertaken which may feed directly into an actual design or production issue.

A suitable project must contain elements of either superyacht design or build technology and include a research element. The types of project undertaken can cover a wide topic range.

Typical projects which could be considered as suitable include:

- A new superyacht platform targeted towards future market sectors an industry partner may be considering, but not have time to work up in house.
- The deployment of a new 'green' hybrid powertrain whether for motor or sailing yachts.
- Developing a low wash hull using SSU's test tank facility which could give a useful insight for both the student and company.
- Improving the production methodology of superyacht build projects.

This only covers a few areas. The fundamental principles of the project are that, ideally, the project should push the boundaries in an applicable area as this will provide the research component necessary for a master's level thesis.

It is not necessary to design the entire vessel. If the project is extensively research or design engineering based, it will be possible to reduce the global outline design aspect of the project.

In simple terms, the scope of an individual project is dictated by the requirement of the sponsoring company providing it is agreed between the university, the student, and the company.

As a general rule, the work completed by the student and the IP rights remain the property of the sponsoring company.

However, the student's thesis needs to be assessed by the university tutor. Once this is completed the work is not published by the university, thereby respecting the commercial rights and integrity of the company's investment.

Our first four MSc students found excellent placement opportunities within the heart of both the UK and European superyacht design and build community during the summer of 2019. The students made good progress within their placements and were able to produce some useful work and valid points for their thesis.

Concluding, one comment from a participating company included: 'The depth of research for the thesis was very good and is of use for ABC Yachting to determine further steps for their building facilities'.

It is hoped that similar placements and comments from industry become commonplace and successful over the next few years as both the university staff and superyacht industry gains experience in finding suitable projects and partners to meet the needs of students and industry alike.

5. CONCLUSIONS

5.1 Summarising the First Year of Operation

The MSc Superyacht Design degree course is new. Therefore it is too early to reflect on whether the course will be successful in the long term. The first intake consisted of four students, some of whom are now employed in the Superyacht industry. This year the number enrolled has grown to seven students. The university's recruitment department is reporting larger numbers showing an interest in the course for next year. This is because, as time passes, the course is becoming known to a wider audience.

Feedback from last year's graduates include the following comments: "What I definitely gained most from the master's degree is the more practical approach. This helped implement my theoretical background into everyday problems. Also, the master's brings the many disciplines of superyacht design together, so I understand the market and role of the designer a lot better than before starting the course".

Students in general were able to join up their previous academic backgrounds with the overall global nature of superyacht design and really gain an overall feel for how their specialism, which may be used in their day to day design specialism, feeds into the final design of the vessel, and how design decisions pertinent to their area will affect other design factors.

Students saw the benefit of their placement and it is likely that one student will be employed by their placement provider. Another has gained employment in the industry as a Junior Superyacht designer and a third is hoping to join a company within the next few months after taking a break.

From the university perspective, the MSc Superyacht Design degree program hopes to gain an international reputation similar to that enjoyed by the two undergraduate degrees. This program is showing potential and, given time, this could be the case.

6. ACKNOWLEDGEMENTS

I would like to thank and acknowledge Solent University's course teaching team for developing excellent teaching material and the first and second cohort for presenting some good coursework, some of which is in this paper.

7. AUTHOR'S BIOGRAPHY

Giles Barkley is Senior Lecturer at Solent University. He is course leader for the three Yacht Engineering degrees and has written the syllabus for the MSc Superyacht Design course. He lectures on all three yacht engineering degrees concentrating on the areas of Design and Naval Architecture.

NAVANTIA'S SMART MAINTENANCE SOLUTION

J. Pagan, NAVANTIA SA SME, Spain

SUMMARY

Maintenance is an essential part of Through Life Support (TLS) service. Some of the most complete concepts for a vehicle sustainment are Condition Based Monitoring Plus (CBM+) and Integrated Vehicle Health Management (IVHM), which use a combination of Condition Based Monitoring (CBM), Reliability Centered Maintenance (RCM) and Artificial Intelligence (AI). Navantia smart maintenance solution integrates CBM+ functionalities in a digital platform to detect anomalies on board and on shore at an early stage. This system provides recommendations to crew; data are also analysed onshore to optimize operation and sustainment management. Ship health status in real time allows the selection of the best ship for a mission, the carrying out of adjustments or repairs before departure, or the moving forward of maintenance tasks. Results show a high capability to detect anomalies with enough response time to avoid catastrophic failures. This current digital platform is compatible with Navantia Integrated Platform Management System (IPMS), and is soon going to evolve into a ship digital twin, planned to be ready for the next generation of naval applications.

NOMENCLATURE

AI	Artificial Intelligence
CBM	Condition Based Monitoring
CBM+	Condition Based Monitoring Plus
CMMS	Centralized Computerized Maintenance Management System
HW	Hardware
ICAS	US Navy's Integrated Condition Assessment System
ILS	Integrated Logistic Support
IoT	Internet of Things
IPMS	Integrated Platform Management System
IVHM	Integrated Vehicle Health Maintenance
JSF	Joint Strike Fighter
RCM	Reliability Centered Maintenance
RUL	Remaining Useful Life
SW	Software
TLS	Through Life Support

1. INTRODUCTION

Key concepts to provide the best support service for a ship throughout its life cycle are reliability and cost-effectiveness. Optimizing reliability allows you to guarantee the best availability rates and to minimize unexpected breakdowns that often result in high repair costs. However, a high reliability TLS service must also consider the concept of cost-effectiveness. Thus, depending on the consequences of a failure, it will be optimal to design a predictive, preventive, or corrective maintenance task.

Reliability must be considered from the very beginning when designing or selecting industrial assets. A good design is essential to achieve a high reliability platform. However, no matter how good the design is, assets deteriorate over time since they operate under a certain load or stress in a real environment, often having a certain randomness. In this way, maintenance is used as an effective way to ensure a satisfactory level of reliability over the life of a ship.

The origin of maintenance was basically to repair after failure: corrective or unplanned maintenance. After that it was mainly time-based: planned maintenance, which establishes periodic intervals to perform the maintenance task. Nowadays assets have become more and more complex and demand greater quality and reliability. This has increased the cost of preventive maintenance and turned it into an excessive expense for the platform. For this reason, more efficient maintenance approaches based on asset condition monitoring have emerged [1].

The most widely used method to maximize reliability is RCM [2] and [3]. It was created for the aeronautical sector in the 1960s [4] and extended and normalized in the 1990s for the defense sector [5, 6] due to its effectiveness. Finally, it was standardized in industry with version RCM2 [7] and RCM3 [8]. RCM is analysed in detail by Rausand [9], where the author demonstrates its usefulness for improving reliability. An example of its application for wind turbines can be seen in [10], while in [11] it is used to plan the overhaul of marine diesel engines. One of the most important features of RCM is that it prioritizes CBM over preventive and corrective maintenance when designing a maintenance strategy.

CBM is a maintenance strategy that recommends repair or replacement actions based on the information collected through the monitoring of asset condition. Its objective is to increase the reliability of assets as well as trying to avoid unnecessary maintenance tasks by performing actions only when there is evidence of abnormal asset behaviour. The architecture and functionalities of a good CBM system must be defined by international standards [12].

In complex assets and systems, a traditional CBM installation, based on vibration signals, is useful but insufficient to detect all failure modes that may occur. This is because vibrations are a robust indicator of mechanical problems but are not able to detect many other types of failure. Therefore, CBM systems have evolved to a data fusion and data analysis systems using much more types of signals, machine/deep learning, and knowledge-based fault databases. A denomination recently used to name the current concept of condition monitoring system is CBM+ [13]. CBM+ combines an integration of multiple sensors with AI techniques and reliability analysis. This concept was created by the United States Department of Defense to improve the reliability and availability of its fleet during the life cycle. Another, even more complete, denomination is IVHM. An IVHM system is a comprehensive system concept that provides automatic support for decision-making about the operation and maintenance of a vehicle. IVHM includes maintenance policy, taking into account the management of spare parts, labour repair facilities, etc. IVHM concept was created more than two decades ago by the aerospace industry [14], where this integrating concept of vehicle health management was born.

At present, the concept of smart maintenance is used for all types of vehicles, including ships, especially in defense. Furthermore, new functionalities are now possible thanks to available disruptive technologies 4.0.

2. SMART MAINTENANCE FOR VEHICLES

An IVHM system uses historical data relevant to the present and future behaviour of the vehicle and transforms it into information to facilitate operational and maintenance decisions. This concept consists of the integration of sensors, communications technologies, asset knowledge, and artificial intelligence to provide the ability to diagnose problems of all vehicle systems and assets and to make recommendations.

The concept of IVHM is an evolution of diagnostic and prognosis systems applied to a traditional CBM system [15, 16, 17]. The objective is to implement an advanced diagnosis, prognosis, and health management strategy that allows continuous monitoring and real-time evaluation of the functional health of the vehicle, predicting the remaining useful life (RUL) of the components that fail or are about to fail, and to use this information to improve decisions about operation and maintenance. This system makes possible an early identification of failures, which results in a reduction of maintenance operations due to a lower number of unexpected failure cases. In addition, when it comes to applications for military operations, the command and control functions may have a greater knowledge of the condition of the vehicle and its capabilities in each moment and situation, which increases safety and chances of mission success. It is also possible to know at the beginning if a vehicle change or timely withdrawal is necessary before a mission begins [18, 19].

For some authors, the scope of an IVHM system also includes logistics management. This means that the availability of vehicle health information in an updated and detailed way can be used to automatically activate logistic actions. In this case, an IVHM system also transmits vehicle status data to a distributed information system that processes IVHM calculations along with other information about the vehicle and the logistics cycle. If maintenance is required, it informs the supply chain infrastructure and support facilities of the need for spare parts, tools, and labour. An example of an expanded IVHM system that includes both the vehicle and its logistic support infrastructure is the Joint Strike Fighter (JSF) program [20, 21] of the United States Department of Defense. JSF program leads the development of attack weapon systems for the United States Navy, Air Force, and Marine Corps. Here, vehicle condition monitoring system is designed with a fully functional IVHM system that performs fault detection, isolation, and reconfiguration through numerous components and subsystems (i.e. structures, engines, electronics, hydraulics, fuel, and electrical systems). The detection of an anomaly that requires a maintenance action activates a logistics chain to schedule the repair or replacement, purchase of spare parts, etc. Currently, there are not many examples of IVHM applications although in recent times initiatives are appearing in all types of aerial, marine, and land vehicles.

Baroth et al. [22] give an overview of the main applications in the automotive, space, defense, and commercial aeronautics sectors. Janasak and Beshears [17] discuss health monitoring systems currently offered to different types of platforms, including vehicles and industrial plants. Similarly, Reichard et al. [23] provide a list of military projects in which the information generated by a vehicle health monitoring system is being incorporated into the automated logistics and strategic fleet management systems. A significative example in the military naval industry is the US Navy's Integrated Condition Assessment System (ICAS) [24] successfully applied to a fleet of more than 100 ships.

The current vision is that new and old vehicles must have smart functions based on advanced technology that allows decisions to be made based on having more information of the design, use, maintenance and logistic support [22, 25, 26]. This is totally consistent with the adoption of IVHM philosophy that uses the fusion and strong coupling of interdisciplinary trends in engineering, computer and communications sciences to achieve the cheapest and most efficient use of systems. An IVHM system consists of three fundamental parts [27]:

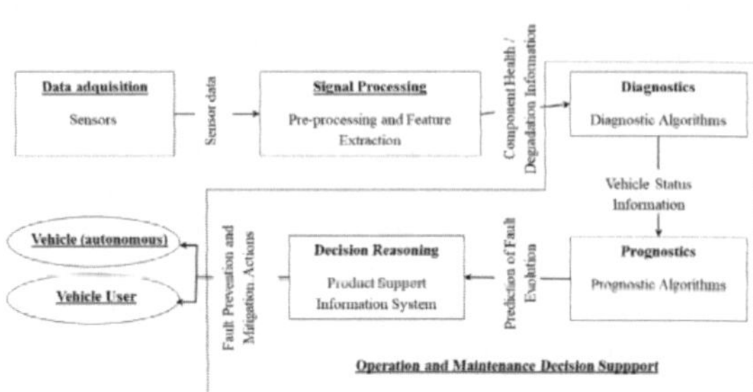

Figure 1: Architecture of an IVHM system

1. Data acquisition. Data collection may be relevant to the system health.
2. Data processing. Management and analysis of the data or signals collected in the previous step for a better understanding and interpretation of the data.
3. Support for operational and maintenance decision making. That includes the tasks of diagnosing and knowing the state of the vehicle, predicting possible failures or evolution of the vehicle and, finally, recommending efficient operation and / or maintenance actions.

Diagnosis and prognosis are two fundamental elements in an IVHM system. Diagnostic tools deal with the detection, isolation, and identification of faults when they occur. Failure detection is the task that indicates if something is wrong with the system being monitored; fault isolation identifies the faulty component; and fault identification finds out the nature of the detected fault. On the other hand, prognosis tools deal with the prediction of failures before they occur. This task determines if a fault is imminent and estimates when and how likely it is to occur. An IVHM system can use diagnosis, prognosis, or both.

However, whatever the objective, it always contains the three key elements of Figure 1.

The literature on diagnosis and prognosis of failures is huge and diverse, mainly due to the wide variety of systems, assets and components existing in the industry. Hundreds of articles in this area, including theories and practical applications, appear every year in academic journals, conference proceedings, and technical reports.

3. SMART MAINTENANCE FOR WARSHIPS

The digital era is revolutionizing maintenance in naval industry due to the availability of big data and artificial intelligence tools at an affordable price capable of analysing and transforming them into useful information to know the health status of on-board assets. Therefore, an already existing IVHM concept only applied in a few military naval existing programs is now expanding to the whole industry.

Among the new emerging technologies 4.0 [28], Figure 2 identifies those that clearly apply to smart maintenance: IoT, digital platform, virtual and augmented reality, cloud, data analysis, modelling and simulation, cybersecurity, and artificial intelligence.

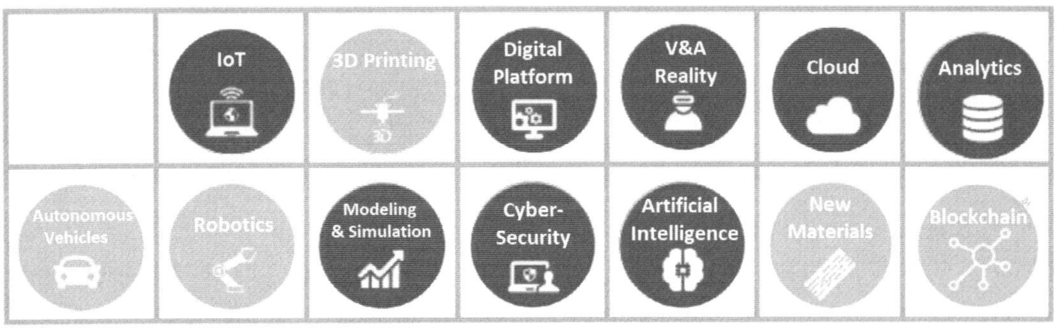

Figure 2: 12 + 1 4.0 technologies applicable to smart maintenance

Basically, a smart maintenance system consists of a data acquisition system, a software (SW) and hardware (HW) infrastructure prepared to monitor and analyse large amounts of data in real time, and a SW application for predicting the health of on-board systems and assets. The platform must have a design that allows interoperability and total scalability to allow the growth of the system as well as adequate security systems that prevent unwanted intrusions. Particularly in the defense sector cybersecurity is a decisive technology to be implemented to allow data transmission. In addition, the concept of smart maintenance is an integrative system. Therefore, interoperability and flexibility are very important factors to allow the integration of signals from specific systems of asset original manufacturers. This allows fault indicators of different nature and technology (operational parameters, fluids, vibrations, thermography, ultrasound, etc.) on the same system.

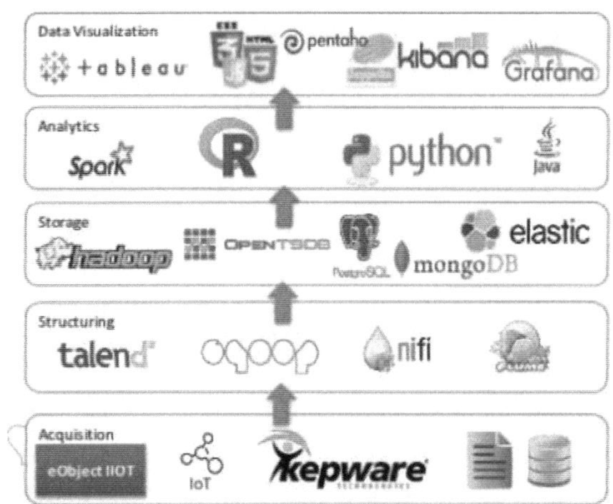

Figure 3: Example of layered architecture of a 'Big Data' and 'Analytics' system

In addition to the capabilities of big data infrastructure, a smart maintenance system must integrate three fundamental pillars into its design: RCM, CBM, and IA. This harmonized combination of knowledge implemented on a scalable, open, flexible, and secure big data infrastructure, will ensure that the solution provides the integration of all available signals that can be transformed into health indicators of assets and therefore allow the prediction and diagnosis of faults, providing adequate information to help the user make decisions regarding operation and maintenance.

Figure 3 shows an update of IVHM concept divided into three blocks: data acquisition, prediction and decision-making support, and maintenance management. The last block corresponds to the integration of a Centralized Computerized Maintenance Management System (CMMS) that allows the management of maintenance actions identified by the prediction and decision-making system. Big Data and AI may also be used among last block CMMS to optimize management of maintenance actions considering maintenance policy and context.

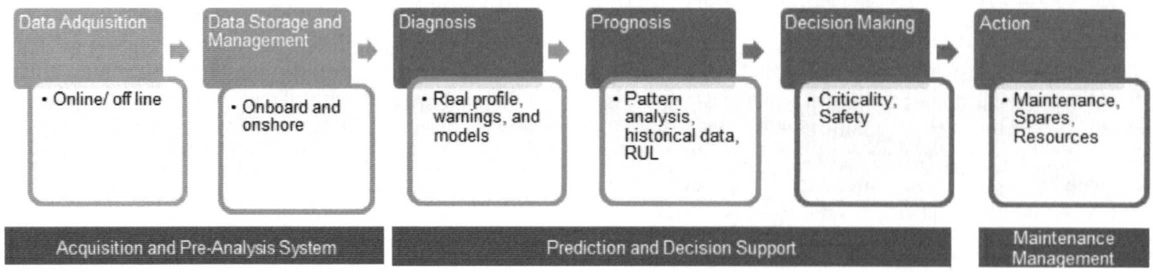

Fig 4. Modules of a smart maintenance system

An example of smart maintenance system is Argos21, developed by Navantia. This system may be installed on board, on shore, and in laboratories. On board and on shore both use the same version: they only have different user profiles inside. However, the laboratory version is designed just for entering the results of fluid analysis and the sending of them to the on shore or on board versions. On the one hand, Argos21 uses the same SW infrastructure as IPMS, so integration on board is perfect; it is possible to install it in the control room and use it in the same IPMS cabinets. On the other hand, this SW infrastructure has evolved to a big data platform so it is possible to connect to everything. This compatibility is especially important on shore where any commercial or open source data analytics or dashboarding tool may be used. Argos21 must be set up at the beginning by specialists, creating tailored normality models of systems and assets using real historical data, and introducing failure data bases.

Depending on the criticality of the system or asset and the available signals, a different strategy may be used for fault detection. If an asset is not critical or there are no enough signals to make a model, tailored limit values are established. Nevertheless, when an asset is critical and the necessary signals are available, a behaviour model is made. In the latter case, if a physical knowledge is available, a knowledge-based model is used and adjusted using real data [29] and [30]. This is the optimal model, from which maximum efficiency could be obtained. If no physical model is known, a data-based model is performed [31]. In this case, a greater number of historical data is needed and false alarms may occur when the vessel sails in an environment other than that used to train the model.

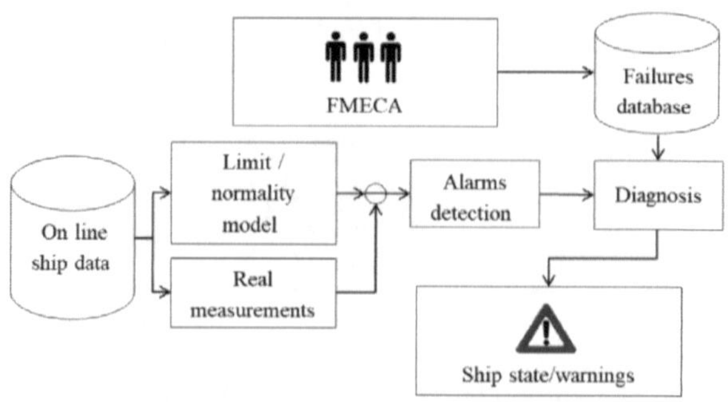

Figure 5: Detection and diagnostic methodology [29]

An expert must construct the ship breakdown structure, where main features, signals, tailored models, limits, and failure database is set up. Moreover, they build layout containing graphs and tools showing key indicators and important parameters to monitor performance and health. Once it has been configured and validated, this system works automatically and is simple to use. It detects anomalies, diagnoses what the incipient failure mode when symptoms appear, and helps decision-making to launch an inspection or maintenance task or to delay a previously scheduled one. The crew profile activated on board shows only the most important information to know in real time ship key indicators and pre-alarms. These pre-alarms or warnings are given by tailored models or limits and include important information to make correct decisions. On the other hand, the expert profile activated on shore shows more complex tools and layouts for advance analyses as, for example, a comparison between ships or maintenance plan analyses. To do this, on shore system configuration contains information about the complete fleet.

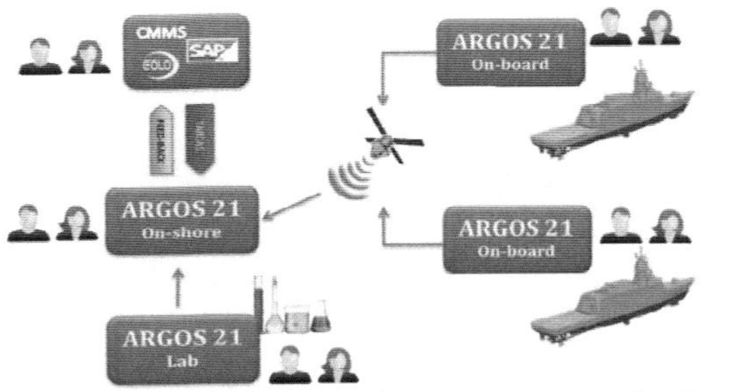

Figure 6: Smart maintenance communications among on board and on shore systems [32]

Figure 6 shows an example of a fleet health status. This information comes from key indicators calculated from models/limits of assets and aggregated to systems and ships. The main objective in warships is to get maximum reliability and availability to perform the mission. For this reason, ship health status combined with other contextual information allows decisions to be made, for example, the selection of the best ship for a mission, the carrying out of adjustments or repairs before departure, or the moving forward of non-critical maintenance tasks.

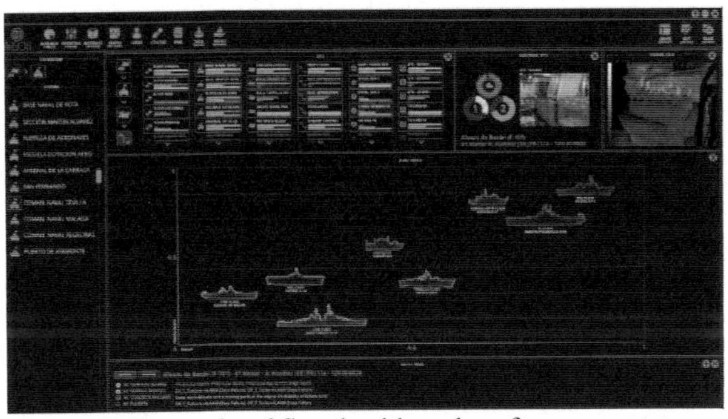

Figure 7: Example of fleet health and performance status [32]

Other applications that Navantia is developing for TLS service using 4.0 technologies are related to maintenance execution and training. Among many applications we may highlight the digitalization of inspection and maintenance tasks using industrial tablets, maintenance training using Virtual Reality, or maintenance assistance using Augmented Reality. Using a complete 3D model of the ship, Virtual Reality can be used as a training platform for maintainers in a fully virtual immersion. Figure 8 shows an example of how to disassemble and clean a main engine oil filter inside the machine room using Virtual Reality. Furthermore, Augmented Reality is a powerful technology to have instant support for performing maintenance tasks at the same time you are doing it thanks to an interactive experience of a real-world environment where assets that reside in the real world are enhanced by computer-generated perceptual information. All these applications reduce time and improve reliability and safety in maintenance execution.

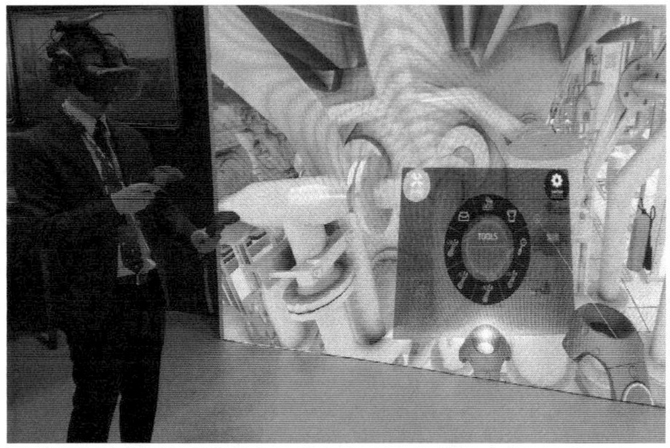

Figure 8: Example of maintenance task training using virtual reality [33].

4 VALIDATION OF SMART MAINTENANCE FOR WARSHIPS - A REAL CASE OF BREAKDOWN

Argos21 demonstrator was built in 2018 to validate the performance of detection, diagnosis, and decision-making support using real historical operating data. Data used for this validation came from both real faults and simulated ones, as there are not real data for all faults of interest. Simulations also use real historical data but symptoms are introduced in a realistic way. Examples of validation cases among others are failures in Main Engines, Generator Sets, Propulsion Shaft Lines, or Turbines. The demonstrator enters data in real time or at a higher speed into Argos21 system and it is checked out when and what it is detected, as well as what is diagnosed and recommended. This paper presents an example of a real catastrophic failure that occurred in a patrol vessel during a mission when there was not a smart maintenance system.

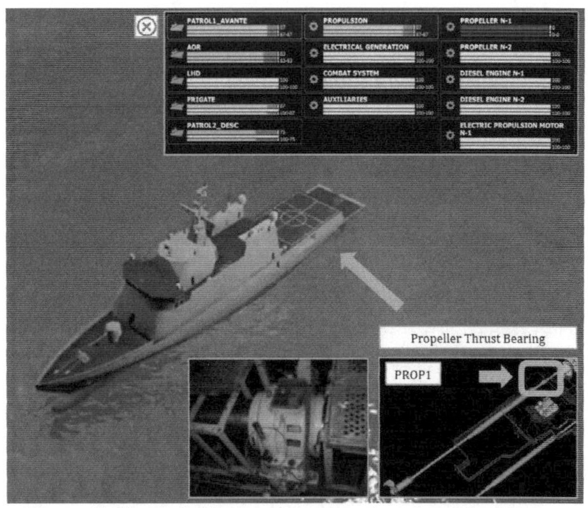

Figure 8: Smart Maintenance example - thrust bearing breakdown (1 of 2)

Here, Argos21 detects a progressive increase in the thrust bearing temperature above the usual value two days before the catastrophic failure occurred.

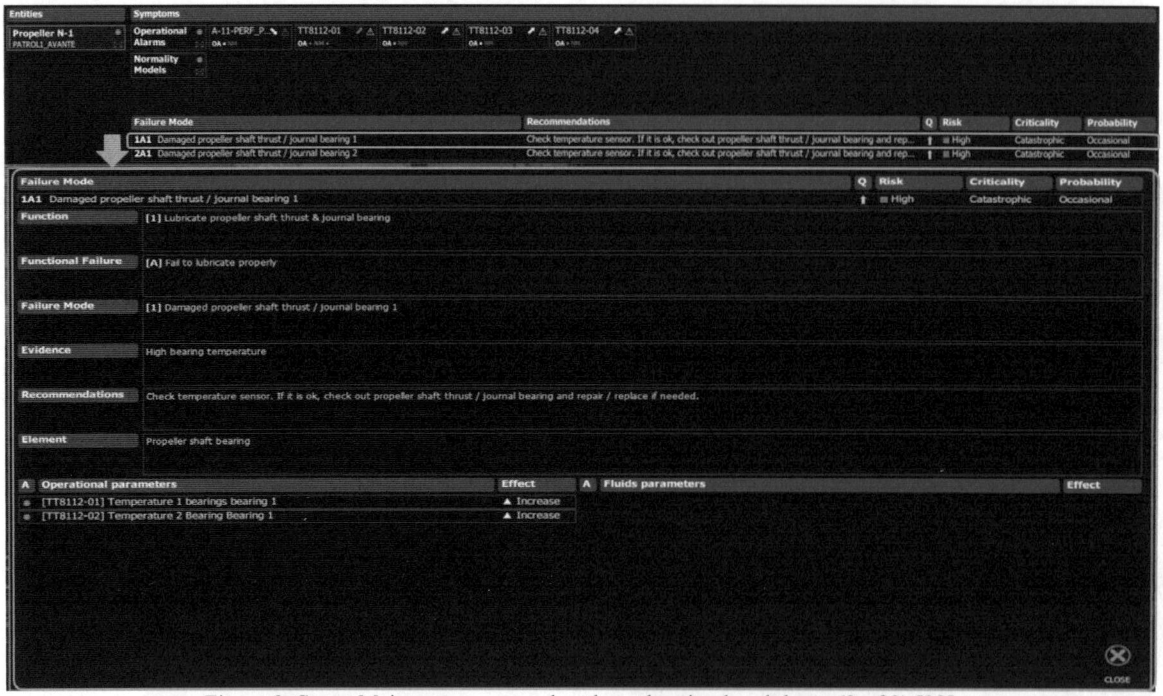

Figure 9: Smart Maintenance example - thrust bearing breakdown (2 of 2) [32]

In this example, Argos21 recommends checking the sensor and whether it is OK to repair the thrust bearing. Moreover, it allows the user to check the diagnosis by displaying parameters of interest related to the fault and to send an automatic report with all related information to send to an expert and/or superior if necessary for decision-making. This example demonstrates that the loss of a propulsion shaft line during a mission as well as the high repair cost could have been avoided. This conclusion may be extended to the rest of validation cases.

5. FUTURE DEVELOPMENT

Navantia's Smart Maintenance solution continues evolving to integrate all current and future applications into the next future digital twin of the ship. To do so, it is necessary to use a digital platform where all data and tools for ship design and through life support are fully integrated. For this reason, it is planned to integrate the current solution on the Siemens platform. This digital platform has been considered the most complete option to use as big data SW infrastructure since it contains everything needed for management, design, testing, modeling, data analysis, maintenance, and other functionalities that comply with all the requisites for shipbuilding and digital twin. Siemens digital platform may also be connected to any commercial or open source data analytics or dashboarding tool. This compatibility is especially important for TLS services when the Navy uses already other maintenance management applications.

6. CONCLUSIONS

This paper contextualizes the role of maintenance within a TLS service of warships. The evolution of maintenance is also shown, especially the integral concepts of CBM+ and IVHM used in the military and aerospace industry. In addition, these concepts are updated to the current context of the disruptive explosion of 4.0 technologies and identified as Smart Maintenance. Finally, it is contextualized in the military naval industry, where Navantia has launched a service called Argos21. Navantia as a designer, builder, and system integrator has the knowledge to create a reliable Smart Maintenance solution adapted to the needs of the Navy.

There are three most important conclusions about Smart Maintenance. Firstly, an efficient smart maintenance system needs to have both an adequate big data SW/HW infrastructure and know-how of maintenance business (maintenance management, reliability, condition monitoring, data analysis) that adequately merges RCM, CBM and IA. Secondly, a smart maintenance system is a key tool that enables the offering of a high value TLS service where reliability and availability can be optimized, as well as costs can be minimized. In this sense, Navies are increasingly demanding this kind of tools that allows them to take advantage of their data in order to know the real state of the fleet at any time, to plan missions and to manage asset maintenance in the most efficient way. And thirdly, it is not easy to provide a TLS service in the military naval sector because of security restrictions. Currently, they do not usually allow data communications outside their intranet. This prevents them from taking advantage of the cloud service. Therefore, Cybersecurity and Blockchain technologies will play a crucial role in the coming years.

On the other hand, the most important conclusions about the Navantia solution are also mainly three. Firstly, Argos21 includes the three smart maintenance pillars RCM, CBM, and AI, since it monitors all signals of interest for health available onboard, generates models, and limits necessary to detect incipient symptoms thanks to AI, and has the knowledge about failure modes thanks to RCM methodology. Secondly, it is built on the same SW infrastructure as the monitoring and control system of ships built by Navantia, which makes it not only fully compatible, but it can be installed as a 'plug and play' system. It is only requires one initial configuration of systems or assets based on real historical data. And thirdly, SW infrastructure is a big data platform with all the features of operability, scalability, and connectivity of these types of platforms. In the near future the integration of the SW infrastructure into the Siemens digital solution to have a complete digital twin of the ship in future Navantia programs.

7. ACKNOWLEDGEMENTS

Special thanks to Navantia TLS business and digital services for their leadership as well as Navantia Engine Factory and Control Systems for their cooperation. Furthermore, thanks to the Universidad Politécnica de Cartagena for the inspiration and training in the study of the state of the art.

8. REFERENCES

1. Martin K.F., 'A review by discussion of Condition Monitoring and Fault-Diagnosis in machine-tools', *International Journal of Machine Tools and Manufacture 34527–551*, 1994.
2. JA1011_200908, 'Evaluation Criteria for Reliability-Centered Maintenance (RCM) Processes', *SAE International, DOI: https://doi.org/10.4271/JA1011_200908*, 2009.
3. JA1012_200201, 'A Guide to the Reliability-Centered Maintenance (RCM) Standard', *SAE International, DOI: https://doi.org/10.4271/JA1012_200201*, 2011.
4. Reliability Web, 'El camino hacia el RCM', *http://reliabilityweb.com/sp/articles/entry/el-camino-hacia-el-rcm/*, 2019.
5. NAVAIR 00-25-403, 'The naval aviation. Reliability-Centered Maintenance', *Management manual process, Direction of commander, naval air systems command*, 2016.
6. NES45, 'Requirements for the application of Reliability-Centered Maintenance Techniques to HM Ships, Submarines, Royal Fleet Auxiliaries and other Naval Auxiliary Vessels', *Defence Standard 02-45 Issue*, 2000.
7. Moubray J., 'RCM2 Reliability-Centered Maintenance', *ISBN-13: 978-0750633581*, 1999.
8. The Aladon Network, 'RCM3 Reliability-Centered Maintenance', http://www.thealadonnetwork.com/wp-content/uploads/2015/12/IntroductiontoAladonRCM3Brochure.pdf, 2015.
9. Rausand M., 'Reliability Centered Maintenance', *Reliability Engineering & System Safety, Volume 60, Issue 2, May 1998, Pages 121-132,* https://doi.org/10.1016/S0951-8320(98)83005-6, 1998.
10. Igba J. et al., 'A Systems Approach Towards Reliability-Centered Maintenance (RCM) of Wind Turbines', *Conference on Systems Engineering Research (CSER'13), Procedia Computer Science 16 814 – 8232013,* https://doi.org/10.1016/j.procs.2013.01.085, 2013.
11. Milkie, C. M. and Perakis, A. N., 'Statistical Methods for Planning Diesel Engine Overhauls in the U. S. Coast Guard', *Naval Engineers Journal, 116: 31–42, 2004.* https://doi.org/10.1111/j.1559-3584.2004.tb00266.x., 2004.
12. ISO 13374-1:2003, 'Condition Monitoring and Diagnostics of Machines -- Data processing, communication and presentation -- Part 1: General guidelines', *https://www.iso.org/standard/21832.html*, 2003.
13. Pack D., 'Fleet management with CBM+', *Army Sustainment magazine. USA,* https://www.army.mil/article/120896, 2014.
14. NASA, 'Research and technology goals and objectives for Integrated Vehicle Health Management (IVHM)', *Report NASA-CR-192656*, 1992.
15. Barber D. E., 'Shipboard Condition Based Maintenance and Integrated Power System Initiatives', *PhD. thesis, Massachusetts Institute of Technology*, 2011.
16. Williams Z., 'Benefits of IVHM: an analytical approach'. *In Proceedings of the 2006 IEEE Aerospace Conference, Big Sky, Montana, USA, 4–, paper no. 1507*, 2006.
17. Janasak K. M. and Beshears R. R., 'Diagnostics to prognostics – a product availability technology evolution', *In Proceedings of the 2007 Reliability and Maintainability Symposium – RAMS'07, Orlando, Florida, USA, 22–25, pp. 113–118,* 2007.
18. Aaseng G. B., 'Blueprint for an Integrated Vehicle Health Management System', *In*

Proceedings of the 20th Digital Avionics Systems Conference, Daytona Beach, Florida, USA, 14–18, vol. 1, pp. 3.C.1-1–3.C.1-11, 2001.

19. Reichard K. et al., 'Integrated Management of System Health in Space Applications', *In Proceedings of the 2007 Reliability and Maintainability Symposium – RAMS'07, Orlando, Florida, USA, 22–25, pp. 107–112*, 2007.
20. Hess A. et al., 'PHM a Key Enabler for the JSF Autonomic Logistics Support Concept', *In Proceedings of the 2004 IEEE Aerospace Conference, Big Sky, Montana, USA, 6–13, vol. 6, pp. 3543–3550*, 2004.
21. Hess A. and Fila L., 'The Joint Strike Fighter PHM concept: Potential Impact on Aging Aircraft Problems'. *In Proceedings of the 2002 IEEE Aerospace Conference, Big Sky, Montana, USA, 9–16, vol. 6, pp. 3021–3026*, 2002.
22. Baroth E. et al., 'IVHM (Integrated Vehicle Health Management) Techniques for Future Space Vehicles', *In Proceedings of the 37th AIAA/ASME/SAE/ASEE Joint Propulsion Conference*, 2001.
23. Reichard K. et al., 'Integrated Management of System Health in Space Applications', *In Proceedings of the 2007 Reliability and Maintainability Symposium – RAMS'07, Orlando, Florida, USA, 22–25, pp. 107–112*, 2007.
24. Finley B. and Schneider E., 'ICAS: The Centered of Diagnostics and Prognostics for the United States Navy', *Proc. SPIE – Int. Soc. Opt. Eng., 4389, 186–193*, 2001.
25. Asiedu Y. and Gu P., 'Product Life Cycle Cost Analysis: State of the Art Review', *Int. J. Prod. Res., 36(N4), 883–908*, 1998.
26. Vachtsevanos G. et al., 'Intelligent Fault Diagnosis and Prognosis for Engineering Systems', *John Wiley & Sons, New Jersey, USA, ISBN: 978-0-471-72999-0*, 2006.
27. Lee J. et al., 'An Integrated Platform for Diagnostics, Prognostics and Maintenance Optimization', *The IMS 2004 International Conference on Advances in Maintenance and in Modelling, Simulation and Intelligent Monitoring of Degradations, Arles, France*, 2004.
28. European Parliament, 'Industry 4.0 - Digitalisation for productivity and growth', *http://www.europarl.europa.eu/RegData/etudes/BRIE/2015/568337/EPRS_BRI(2015)568337_EN.pdf*, 2015.
29. Pagan J., 'Sistema de Diagnosis de Motor Diesel Marino Basado en Modelo Termodinámico e Inteligencia Artificial', *PhD. thesis, Universidad Politécnica de Cartagena, https://www.researchgate.net/profile/F_Vera-Garcia/publication/318674337_Sistema_de_diagnostico_de_motor_diesel_marino_basado_en_modelo_termodinamico_y_de_inteligencia_artificial/links/59772ed2458515e26d2cde9f/Sistema-de-diagnostico-de-motor-diesel-marino-basado-en-modelo-termodinamico-y-de-inteligencia-artificial.pdf*, 2017.
30. Pagan J. et al., 'Marine Diesel Engine Failure Simulator based on Thermodynamic Model', *Applied Thermal Engineering, Volume 144, 5 November 2018, Pages 982-995*, https://doi.org/10.1016/j.applthermaleng.2018.08.096, 2018.
31. Diez-Olivan A. et al., 'Data-Driven Prognostics using a Combination of Constrained K-Means Clustering, Fuzzy Modelling and LOF-Based Score', *Neurocomputing, http://dx.doi.org/10.1016/j.neucom.2017.02.024*, 2017.
32. Pagan J., 'Argos21 – Smart Maintenance System for Navantia Ships'. *Internal Navantia R&D project. Unpublished work*, 2019.
33. Muñoz J.L., 'NAVANTIS -, 'Navantia digital platform for operation and maintenance training'. *Internal Navantia R&D project. Unpublished work*, 2019.

9. AUTHOR'S BIOGRAPHY

J. Pagan works in Smart Maintenance at Navantia. He is responsible for the co-ordination of R&D projects related to the integration of 4.0 technologies for TLS services in Navantia. His previous experience includes Integrated Logistic Support (ILS), predictive maintenance, condition monitoring, and R&D coordination. He won the National PhD Thesis Award in 2017 given by the Spanish Maintenance Society, and was finalist of the European PhD Thesis Award in 2018 given by the European Federation of National Maintenance Societies.

LASER HYBRID WELDING IN SHIPBUILDING

C. Churiaque, M. Chludzinski, R.E. Dos Santos, J.M. Sánchez-Amaya, University of Cádiz, Spain
M. Porrúa-Lara, F. Abad-Fraga, Navantia S.A., S.M.E. Astillero Bahía de Cádiz, Spain

SUMMARY

The progress of the shipbuilding industry 4.0 involves addressing a number of challenges in shipyards for the development and improvement of more efficient and productive processes. The joining of flat plates is one of the first stages in the manufacturing process of large vessels which is pivotal to the production of the whole ship. High Power Laser Arc Welding (HLAW) is a deep penetration and highly efficient welding process that produces welds with a high depth to width ratio, providing high quality joints. This paper reviews the most influential process variables in thick plate naval steel butt welds and gives a brief summary of techniques applied to improve the quality of the joints and key characteristics of the equipment used. Recent published studies show good results in butt welds of high thickness (28mm) in single pass using a magnetic backing. The use of HLAW achieves a significant reduction of seam irregularities, which can act as potential starting points of failure during service. The reduction in distortion, coupled with reduced bead size and increased quality and speed, make hybrid laser welding a realistic alternative of the conventional submerged arc welding. From this study, it seems that a single pass procedure to ensure high quality of thick welds is one of the key aspects required by industry, to be covered by the near future research. The present paper reviews the state-of-the-art of the Hybrid Laser-Arc Welding process applied to shipbuilding industry.

Keywords: hybrid laser arc welding; thick thickness; butt weld; naval steel; shipbuilding

1. INTRODUCTION

In the Hybrid Laser-Arc Welding process (HLAW), the laser beam and electric arc interact together in the weld pool and their synergistic effect is used to perform deeper and narrow welds, increasing the production rate [1-3]. Laser welding has gained popularity in recent years because laser beam focuses energy in a very small area [4]. The precision and concentration of energy provided by laser beam increases the welding speed, which in turn reduces the heat input and the thermal distortion in the welded parts [5]. However, laser welding has little capacity for bridging or filling the gap and therefore requires great precision in the assembly of workpieces and preparation of edges [6].

Contrary to the singularities of laser technology, arc welding processes have excellent gap filling capacity, high electrical efficiency, and allow the efficient welding of materials with high reflectivity, such as aluminium [7]. Arc welding systems are much cheaper than laser welding systems with similar capabilities [8]. However, the low energy density during arc welding makes the process slower, which induces a high heat input in the welding zone, resulting in thermal distortion of the welded part. Laser and arc processes, when applied together in the same welding pool, provide a hybridization effect that compensates the disadvantages of both processes and enhances its advantages [9].

At the end of the 1970s, Steen et al. [10] introduced the HLAW process. The results of his research clearly showed the advantages of combining a laser beam and an electric arc for welding. The hybridization effect of the plasma arc with a CO_2 laser beam showed a remarkable increase in the welding speed (up to 100%), the weld depth (penetration) and the stability of the process [11]. After the demonstrated success of the application of this technique, the research and development of HLAW underwent a slow growth, due to the lack of availability of high-power lasers. The development of high-power industrial lasers at the end of the 1980s [12] attracted the attention of researchers from the industrial sector and the academic world, who guided their studies towards improving the process to make it viable for industrial use.

HLAW presented a greater development in the nineties due to the availability of CO_2 lasers that could concentrate densities of higher powers, up to 10^6 W/cm^2. Since then, several studies have been focused on materials approach applied on steel [11,13], titanium [14,15], aluminium alloys [16,17], or dissimilar joints [18,19]. Diverse investigations have been carried out to face different aspect of the process, such as the economic viability of the laser-arc hybrid welding process, the influence of the relative position of the laser beam and the torch of the electric arc, edge preparation, minimization of metallurgical defects, and optimization of process parameters, etc [20–23]. However, despite the multiple advantages to the process, its industrial application was limited due to the high cost of the equipment and the complexity of the process, which requires a large number of adjustment parameters.

Another influential factor into the inactive state of this technology was due to the lack of commercial availability of an integrated hybrid laser welding system. The first industrial system of HLAW was introduced in 2000 by the Fraunhofer Institute in Germany, for the manufacture of industrial tanks [24].

The industrial applications of hybrid welding have grown in this century due to the advances of automation and control, as well as the tooling design. This progress has facilitated the development of integrated and commercial hybrid welding heads [8]. Among the companies and partnerships having had succeed in HLAW integration, the following can be cited: ESAB, Fronius, Lincoln Electric, IPG Photonics, Trumpf, Cloos, PEMA, or IMG [25–31]. Currently, laser systems are becoming cheaper, more reliable and efficient and with higher power.

It is worth mentioning that the driving force of the introduction of this technology in the building cruise ships and military vessels is the reduction of the thermal distortion generated by welding, which leads to a reduction in the corrective processes after welding, such as straightening/smoothing operations (whose costs are estimated at around 15–30% of the total staff costs in new constructions) [32].

Another aspect to be seriously taken into account is that the replacement of the conventional welding techniques by HLAW involves not only an increase of the welding speed itself (from 0.22m/min with submerged arc welding, up to 1,3 m/min with HLAW), but also the overall manufacturing production rate of shipyards. A recent simulation study [33] demonstrated that a potential productivity increase of 33% can be gained in terms of numbers of shipyard structures manufactured per year.

Leading shipyards worldwide have already implemented HLAW technology mainly for the construction of flat panels. An active use of the laser welding technologies applied to shipyards is performed by Meyer-Werft (Papenburg, Germany) company, which employs CO_2-laser welding and HLAW at the productions of the cruise liner panel of 20 metres length. The Kvaerner Warnow Werft (Rostock, Germany) company is equipped with a portal HLAW system with a fibre/disc laser. Fincantieri Shipyards (Monfalcone, Italy) and STX Finland Cruise Oy (Turku, Finland), are also equipped with laser and HLAW technologies for shipbuilding [34].

Currently, HLAW technology is facing new challenges and the application's interest arises in the building of large vessels. HLAW allows penetration depths in a single pass unreachable with other processes. The concise achievable penetration depends on the characteristics of the laser source used; HLAW usually allows welding relatively high plate thicknesses (in the order of 10 to 15mm) [21]. Hybrid welding is proposed as an ideal joining process for large-scale applications, such as those required in the building of chemical tankers, gas or oil tankers, since the thickness of material to be welded is usually greater than 20mm [35].

2. METHODOLOGY

2.1 MATERIALS

There are several types and grades of naval steels, with different chemical composition and mechanical properties. These characteristics are essential to the selection of type of steel applied to different structures, according to the specific application requirements. Among them, the following can be mentioned: medium steel, high strength alloy, and structural steel. In the same way, there are diverse types of filler wire, whose chemical composition and mechanical properties are oriented to different categories of steels.

The classification societies have their own rules for the design, construction, and inspection of shipbuilding structures. The key difference between the different grades relies on the temperature in which Charpy tests are performed.

2.1 (a) Base Material

Grade A is a general-purpose mild steel for use in non-critical locations [36]. High strength steels are required for liquefied petroleum gas or methane carriers (in liquid form at atmospheric pressure or at atmospheric temperature under pressure). High tensile quenched and tempered steels are also used in naval construction, particularly for aircraft carrier flight decks, side armour, magazine protection, and the hulls of submarines. Other structural steels of interest for the joining by HLAW process are HSLA (High-strength low-alloy steels) such as: X65, X70, X80 X120 pipeline steel, for offshore industry. An example of this study was carried out by Rethmeier et al. [37], in which the base material selected was the typical pipeline steel API 5L X65. Another application reported in the literature involved the welding lightweight sandwich panels (know as LASCOR) made of duplex stainless steel [38].

2.1 (a) Filler Wire

Frequently, in the HLAW process, the laser is used to achieve the maximum penetration possible, and the arc welding, for the incorporation of additional material that allows for the covering of gaps and/or to modify the composition of the material in the welding zone [11, 35]. In this sense, the contribution of a filling material is particularly advantageous for large scale applications.

Studies carried out in [39] indicate that, using the parameters suitable for the hybrid process, the tolerance of the spacing between sheets can be increased by a factor of three or four, depending on the materials and equipment used, if compared with autogenous laser welding [39, 40]. In Reference [32], it is indicated that the greater benefits of hybrid welding compared to autogenous laser welding is the increase in the tolerance of the spacing between workpieces, being able to execute a sound weld with spacing in the order of 1 mm [39]. Table 1 presents some of the main filler wires used in the bibliography for the HLAW applied to thick steel plates.

Table 1. Filler wires used in HLAW for different steels.

Base Metal	Filler Wire	Ø (mm)	Ref.
RS E36	Pb 60M	1.6	[39]
S355J2 **API 5L X65**	G3 Nil	1.2 1.6	[37] [39] [40]

S235 **S355** **API 5L X65**	G3 Sil	1.2	[37] [41] [42]
HSLA Steel plates	Trustarc MX-A55T	1.2	[43] [44]
SM490A With cut-wire	MG-50 YM-26 ER70SG	1.0	[45]
Q235 Mild steel	ER70S-6	1	[46]

The addition of filler material also involves changes in the chemical composition, microstructure, and mechanical properties of the joint [44]. In this way, the quality of the welds can be improved, since the appearance of solidification cracks is avoided and the porosity in the welds is minimized [47, 48].

2.2 STANDARDS

The increase of HLAW application is also driving the examination of what is an acceptable hybrid weld. DNV and Lloyd's have already developed standards for hybrid welding for shipbuilding applications. However, with the recent activities for HLAW in other industries, there has been additional interest and the need for the development of other standards and specifications by other organizations. The American Society of Mechanical Engineers (ASME) and the American Welding Society (AWS), both already having laser specifications, are active in the development of standards and recommended practices for hybrid processing. Specifications of both organizations are requested by companies using those specifications to develop their Procedure Qualification Records (PQR) and their Welding Procedure Specifications (WPS). These standards address the parameters needed to be documented and how much variation or change are allowed before approval of partially or fully re-qualified process [49].

The International Organization for Standardization (ISO) creates documents that provide requirements, specifications, guidelines or characteristics that can be used consistently to ensure that materials, products, processes, and services are fit for their purpose [49]. The following Table 2 shows the ISO standards applicable to the HLAW process.

Table 2. HLAW Standards.

Identification	Description
ISO 12932:2013	Welding. Laser-Arc Hybrid Welding of Steels, Nickel and Nickel Alloys. Quality Levels for Imperfections.
ISO 15614-14:2013	Specification and Qualification of Welding Procedures for Metallic Materials. Welding Procedure Test. Part 14: Laser-Arc Hybrid Welding of Steels, Nickel and Nickel Alloys.
ISO 15609-6:2013	Specification and Qualification of Welding Procedures for Metallic Materials. Welding Procedure Specification. Part 6: Laser-Arc Hybrid Welding.
ISO 15613:2004	Specification and Qualification of Welding Procedures for Metallic Materials - Qualification Based on Pre-Production Welding Test.

2.3 HLAW PARAMETERS

The quality of the joints obtained depends on the ratio of the power between the two sources of energy and also on the design parameters of the system. There are many welding parameters to define, as it involves a combination of laser and arc processes. However, the main parameters can be grouped into design parameters (thickness and gap geometry), combined process parameters (welding speed, process distance, power, feed rate), and material parameters (chemical composition of base material and filler wire, mechanical properties), as show in Figure 1. These parameters significantly influence the process result [50].

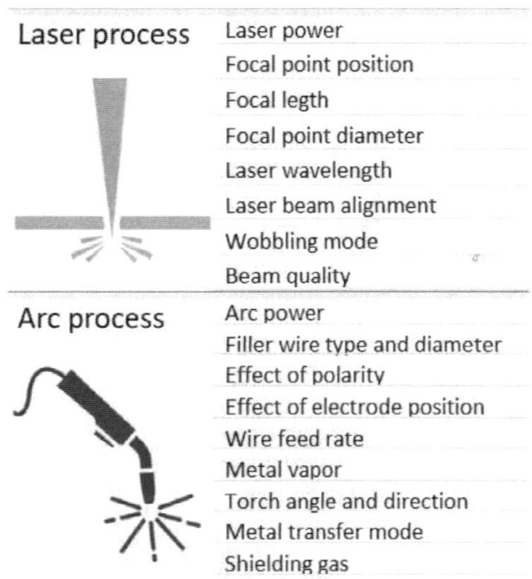

Figure 1: HLAW parameters.

2.4 HLAW CONFIGURATIONS

Another key experimental parameter is the relative position between the laser and arc source, known as laser or arc leading. According to [51, 52], the "Leading Arc" (Trailing Laser) configuration provides a deeper penetration, more stability to the arc process, and lowers the disturbances of the shielding gas, better protecting the arc. "Leading Arc" is also recommended for butt joint configurations without gaps, allowing also higher welding speeds. In this configuration, the arc produces a preheating effect in the joint to be welded that increases the temperature of the material and improves the absorption of the laser [51]. In contrast, "Leading Arc" configuration causes more spatters.

On the other hand, the "Leading Laser" (Trailing Arc) configuration provides more stability to the keyhole, increases the pool size at the upper part of the welding, and improves the de-gassing effect, thus reducing the levels of porosity in the joint. It also allows a wider weld shape and smoother metallurgical transition between the weld and base metal. Besides, Leading Laser is better recommended for bigger welding gaps, presenting greater bridging capacity [52].

2.5 COMBINATIONS OF HLAW PROCESSES

HLAW is a combination of electric arc welding and laser beam welding, working in the same welding group. Fundamental progress is being made with respect to the development of combinations of conventional welding processes, with hybrid laser welding. Some of the basic configurations of application in shipbuilding are presented below.

2.5 (a) HLAW + MIG/MAG

In addition to conventional hybrid laser arc processes with a laser beam as the primary heat source and an electric arc as the secondary heat source (Figure 2), there have also been some technological developments that use two or even more electric arcs.

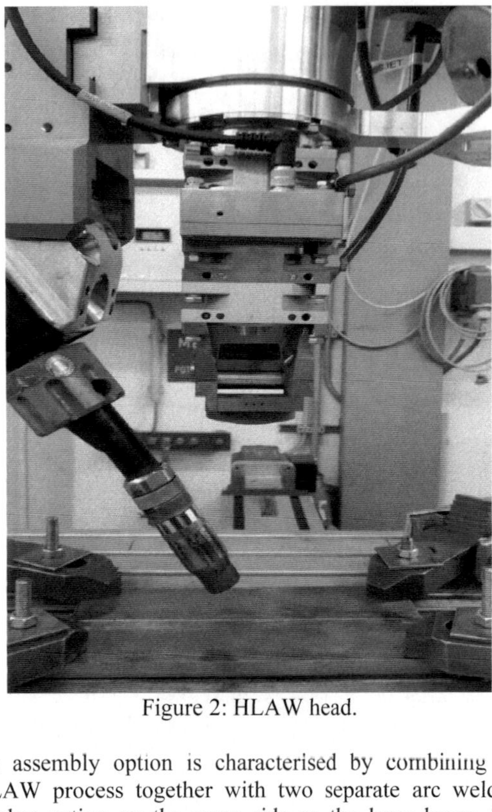

Figure 2: HLAW head.

An assembly option is characterised by combining the HLAW process together with two separate arc welding torches acting on the same side as the laser beam. The HLAW head and weld arcs are aligned along the weld seam with the main arc torch of the head in an arc leading configuration, and the second torch in trailing arc configuration. This type of arrangement, also known as the Hydra process (hybrid double fast arc welding), was originally applied by [53], which combined a CO_2 laser welding process with GMAW processes. The deposition rate of the filler material is considerably increased thanks to this welding combination, which translates into higher welding speed and lower thermal load [53].

2.5 (b) HLAW + SAW

The MIG/MAG laser hybrid welding process encounters problems in some applications with respect to pores at the root of the sheet when welding a plate more than 12 mm thick with partial penetration [53]. The combination of HLAW with the submerged arc process aims to keep the melt bath as long as possible so that the bottom of the seam can be degassed through the seam surface. Submerged arc welding is a high quality, reliable and productive welding process that has been successfully used in shipbuilding for decades. Due to the preheating introduced by the laser welding process, the welding speed of the submerged arc process may slightly increase. The spatial distance between processes, the angles of inclination of the heat sources with respect to the part to be welded, as well as the gap, are the main focus in the investigations carried out so far [54].

2.5 (c) HLAW + CUT-WIRE

Recently, Wahba et al. [45] experimented the combination of HLAW and cut-wire welding processes for joining thick steel plates. The grooved plate placed below the workpieces solved the problem of lack of metal deposition in the root. The cut-wire provided the homogeneity of the filler metal in all its depth providing a significantly greater uniformity in the joint, which substantially improved the deposition of filler material into the gap before welding. Based on this, plates up to 50mm thick were welded using a double pass welding technique with no critical imperfections [44,47].

3. RESULTS

The present section briefly revises the overall concerns and results of the analyses performed to HLAW welds in the literature, including microstructural characterisation and evaluation of properties.

The application of HLAW in shipbuilding is still considered as a challenge. Several investigations are developed using different techniques and experimental methods for the execution of these welds. All researches aim to develop welding procedures producing joints able to pass the industrial qualification processes and the approval required by standards. The influence of processes parameters (for both arc and laser sources), joint configurations and materials play an important role. The aim is to achieve sound welds with complete penetration and without critical imperfections, as shown Figure 3. As a consequence, the investigations perform several evaluations to guarantee the quality of the joints, where the metallurgical characterisation and mechanical properties must be analysed. The quality of joints in terms of macro and microstructure, tensile strength, hardness, impact toughness, and fatigue are studied.

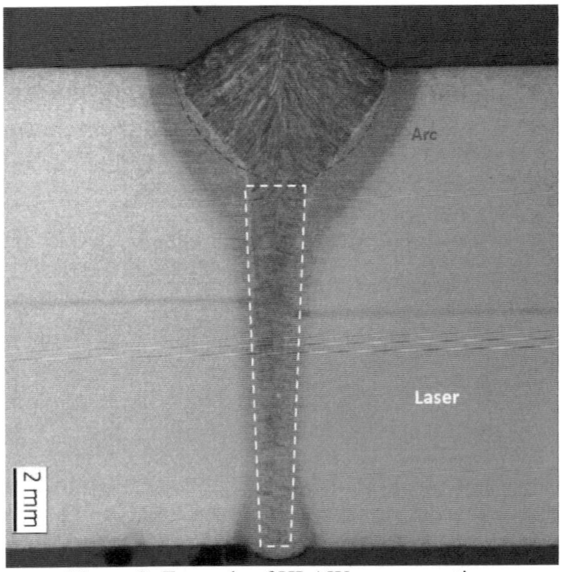

Figure 3: Example of HLAW macro-section.

In addition, the results obtained with the application of the hybrid process lie in the high welding depth.

Research is progressing towards single-pass welds of large thickness plates, or reducing the number of weld layers to the minimum possible in order to minimize thermal distortion and considerably reduce welding time. Likewise, it is of greater interest to minimize the energy during the welding process.

The HLAW technique combined with filling GMA process (instead of saw) in the laser leading configuration caused problems due to greater porosity, since when welding on leading laser, the gas cannot be expelled through the seam root. It is worth mentioning that the hybrid submerged arc laser welding process was derived from these results [54]. Reported results show that plates of 35mm thickness can be welded without problems using the Laser-GMA hybrid welding process as a root layer (18mm welding depth), follow by SAW process to fill up to 35mm. No defects were detected in the joining of flanks and layers.

After executing the welding tests, joint samples are usually characterised. The joints are first visually inspected. Additionally, a full set of Non Destructive Tests (NDT), including X-Ray testing, should be performed. The homogeneity and continuity of the welds are metallographically evaluated according to standards in order to report the weld size, weld morphology, microstructure, and the presence of macroscopic defects.

Subsequently, welds are further evaluated by appropriate mechanical testing. Usually, those welding conditions leading to appropriate joints according to the previous tests (NDTs and microstructure characterisation) are subjected to mechanical tests, such as hardness, bending, and tensile tests. All these mechanical tests are carried out according to ISO standard tests.

3.1 MICROSTRUCTURE

The mechanical properties of a joint are influenced by microstructure promoted during the welding. The arc and laser zones displays distinguish features. The characteristics of the arc process have a considerable and effective role in the welding results.

The chemical composition of the steel and filler wire and especially the carbon content will significantly influence the resulting microstructure. Besides the features of the materials involved, the welding procedure generates a significant modification in the metallurgical characteristics. Each weld zone develops different microstructures, associated with the cooling rates and travel speed. HLAW welds normally have two clearly different fusion zones, so - called laser zone (at the lower part) and arc zone (at the upper part). In general, the microstructure generated at the laser zone is harder (bainite or even martensite), composed of fine grains in the fusion zone (centre line), while the HAZ demonstrated restrain to the grain growth. On the other hand, the microstructure of the arc zone may develop a coarse grain in the fusion zone with wider ferrite proportion in the HAZ [46,55,56].

3.2 MECHANICAL PROPERTIES

3.2 (a) Hardness

Generally, the microhardness of both the arc and laser zones, as well the respective HAZs, are higher than the base material. For industrial applications, harder microstructures must be avoided during the welding in order to prevent the development of cracking mechanisms during service life. Therefore, relevant standards set a maximum limit of 350 HV [55].

3.2 (b) Tensile Strength

The modification promoted by the hybrid process on the tensile strength normally provide sound outcomes, independently of the quantity of layers and technique applied. If the specimens pass the END tests and macro quality, the tensile test is usually not a problem if high heat input values during the process are not achieved. Tensile specimens can be taken from the top, middle, and bottom of the weld seam to test the seam areas affected by the hybrid process. Analysed studies [54, 57] show that all these specimens break in the unaffected base material away from the weld seam, a condition required to pass the specification. This means that the weld itself is not the weakest point of the structure. The minimum required elastic limit is achieved for each part of the weld. Based on the results, it is considered that the tensile test has been passed for the three areas.

3.2 (c) Impact Toughness

The Charpy test is widely applied in industry for analysing the steel joints. Similar to other properties, it is necessary to examine each welding region to define the effect of the welding processes. This test is essential to guarantee the joint quality, as there is always risk of martensite formation in laser welding fusion zone which can lead to impact toughness drop.

3.2 (d) Fatigue

The fatigue behaviour is an important issue for any joint, especially for all cyclic loaded welded structures used any industrial application, such as ships and offshore structures. In thick steel hybrid joint, the concern must be even higher due to the inhomogeneous characteristics developed during the welding [41]. Even with the initial stage of investigations involving this subject, the results suggest good fatigue performance promoted by the hybrid process. Some research found shows that the fatigue strength analyses of the hybrid process provides better performance than joints welded only with the submerged arc process. The results are related to the reduced stress concentration of the flatter weld transition [42]. To date, there are only few studies related to fatigue characteristics applied to hybrid joint of thick steel. Given the high importance of the fatigue behaviour in shipbuilding, future investigation should focus on performing dynamic tests on hybrid welds [58].

4. CONCLUSION

In the shipbuilding industry, welding is considered one of the most critical operations. Until now, electric arc welding has been used as the main process applied for joining structural steels. Nevertheless, the consolidation of the use of HLAW in thicker materials allows it to become a promising alternative for the replacement of conventional processes in shipbuilding. The hybridization compensates for the disadvantages of both processes and enhances its advantages, promoting higher productivity. It is a stable process providing high weld quality and has the possibility of being integrated in an automated production line, reducing the associated costs. Such characteristics are needed in the shipbuilding industry and they are the driving force for the introduction of this technology in this sector.

Different combinations of HLAW process were applied to achieve butt joints with complete penetration and without critical imperfections. Those joints demonstrated, as expected in thick materials, the difference between the microstructures developed by each type of welding processes, according to the steel employed. In general, the laser process inputs a rapid thermal cycle that generates finer and harder microstructure in the fusion zone than the arc process. The HAZ promoted by the arc source displays a higher extension, which is associated with the technique involved. These modifications defined the mechanical properties.

Despite the technological and scientific advances already accomplished, efforts must continue in order to achieve a higher degree of industrial application, especially at shipbuilding or offshore structures. The researches have demonstrated the advantages of the HLAW and promising results in terms of feasibility of the process and joint quality. Joints were performed with high welding speed and suitable fit-up tolerance, showing deep penetration, and high tensile strength and fatigue strength.

From the industrial point of view, the joining process should be executed in a quick and simple procedure in order to maximize the economic benefits in an automated production line. Therefore, future research is needed to improve the HLAW technology in order to be able to weld thick plates in a single pass.

5. ACKNOWLEDGEMENTS

The present research is part of the PhD developed by C. Churiaque, and is funded by the Industrial PhD program of the UNIVERSITY OF CÁDIZ (UCA) and NAVANTIA S.A, S.M.E. and by the JOINT INNOVATION UNIT (UNIDAD DE INNOVACIÓN CONJUNTA, UIC) between UCA and NAVANTIA S.A, S.M.E., co-financed by IDEA Agency of Junta de Andalucía.

Author C. Churiaque would like to thanks to CEIMAR Fundation for funding her research stay at BAM under the "INTERNATIONAL MOBILITY AID TO CENTERS OF EXCELLENCE FOR DOCTORAL STUDENTS CEI·MAR 2019". Also, the grant for the realization of Research Stays in Universities of Prestige (UCA International) is gratefully acknowledged to Vicerrectorado de Política Científica y Tecnológica.

6. REFERENCES

1. Churiaque, C. Laser Hybrid Butt Welding of Large Thickness Naval Steel. *Metals*. 2019.

2. Zhang, S. The technology and welding joint properties of hybrid laser-TIG welding on thick plate. *Opt. Laser Technol*. 2013.

3. Atabaki, M.M. Experimental and numerical investigations of hybrid laser arc welding of aluminum alloys in the thick T-joint configuration. *Opt. Laser Technol.* 2014.

4. Cui, L. Microstructures and Mechanical Properties of Dissimilar Al/Steel Butt Joints Produced by Autogenous Laser Keyhole Welding. *Metals*. 2017.

5. Steen, W.M. Laser Material Processing, *4th ed.; Springer: London, UK*. 2010.

6. Unt, A. Fiber Laser and Hybrid Welding of T-joint in Structural Steels. *Acta Universitatis Lappeenrantaensis 835*. 2018.

7. Guen, E.L. Analysis of hybrid Nd: Yag laser-MAG arc welding processes. *Opt. Laser Technol*. 2011.

8. Ishide, S. Latest MIG, TIG, arc-YAG laser hybrid welding systems for various welding products. *In Proceedings of the International Symposium on High-Power Laser Macroprocessing, Osaka, Japan*. 2002.

9. Ascari, A. The influence of process parameters on porosity formation in Hybrid LASER-GMA welding of AA6082 aluminium ally. *Opt. Laser Technol*. 2012.

10. Steen, W.M.; Eboo, M.; Clarke, J. Arc augmented laser welding. *In Proceedings of the 4th International Conference on Advantages in Welding Processes*. 1978.

11. Górka, J. Microstructure and Properties of Hybrid Laser Arc Welded Joints (Laser Beam-MAG) in Thermo-Mechanical Control Processed S700MC Steel. *Metals*. 2018.

12. Seyffarth, P. Laser-arc Processes and Their Applications in Welding and Material Treatment. *Taylor & Francis: London, UK*. 2002.

13. Turichin, G. Laser-Arc hybrid welding perspective ultra-high strength steels: Influence of the chemical composition of weld metal on microstructure and mechanical properties. *Procedia CIRP*. 2018.

14. Shi, J. Application of Pulsed Laser-TIG Hybrid Heat Source in Root Welding of Thick Plate Titanium Alloys. *Appl. Sci*. 2017.

15. Li, C. Fiber laser-GMA hybrid welding of commercially pure titanium. *Mater. Des*. 2009.

16. Casalino, G. Arc leading versus laser leading in the hybrid welding of aluminium alloy using a fiber laser. *Procedia CIRP*. 2013.

17. Huang, L. Effect of the welding direction on the microstructural characterization in fiber laser-GMAW hybrid welding of 5083 aluminum alloy. *J. Manuf. Process*. 2018.

18. Casalino, G. Effects of Laser Offset and Hybrid Welding on Microstructure and IMC in Fe–Al Dissimilar Welding. *Metals*. 2017.

19. Gao, M. Microstructure and Tensile Behavior of Laser Arc Hybrid Welded Dissimilar Al and Ti Alloys. *Materials*. 2014.

20. Dilthey, U. Expanded capabilities in the welding of aluminium alloys with the laser-MIG hybrid process. *Aluminium*. 1999.

21. Dilthey, U. Prospects by combining and coupling laser beams and arc welding processes. *Riv. Ital. Saldatura*. 1999.

22. Graf, T. Laser Hybrid Process at Volkswagen; IIW-Doc. XII-1730-02. *International Institute of Welding: Copenhagen, Denmark*. 2002.

23. Dilthey, U. Technical and economical advantages by synergies in laser arc hybrid welding. *Weld. World*. 1999.

24. Petring, D. Recent progress and innovative solutions for laser-arc hybrid welding. *In Proceedings of the 1st Pacific International Conference on Application of Lasers and Optics, Melbourne, Australia*. 2004.

25. ESAB. Hybrid Laser Welding. *Available online: https://www.esabna.com/us/en/products/welding-automation-robotics/laser-beam-welding-lbw/hybrio-hybrid-laser-welding.cfm* (accessed on 30 November 2019).

26. Carl Cloos Schweisstechnik GmbH. *Industry 4.0. Available online: https://www.cloos.de/de-en/* (accessed on 30 November 2019).

27. Cloos. Available online: *https://www.cloos.de/de-en/products/qineo/laser-hybrid-weld/* (accessed on 30 November 2019).

28. Lincoln Electric. Hybrid Laser Arc Welding. Available online: *https://www.lincolnelectric.com/assets/US/EN/literature/mc1129.pdf* (accessed on 28 October 2018).

29. Fronius. LASERHYBRID. Available online: *https://www.fronius.com/en/welding-technology/our-expertise/welding-processes/laserhybrid* (accessed on 30 November 2019).

30. PEMA Welding Automation. Pema Welding and Production Automation for Shipbuilding. Available online: *https://pemamek.com/welding-solutions/ship-building/* (accessed on 30 November 2019).

31. IMG Engineering Technology and Mechanical Engineering GmbH. Individually Customized Services and Products for Heavy Industry. Available online: *https://www.img-tech.de/en/services* (accessed on 30 November 2019).

32. Mundada, S.S. Process Improvement through Hybrid Laser-MigWelding in Medium Scale Farm Equipment Industry. *Int. J. Latest Trends Eng. Technol*. 2016.

33. Sánchez-Sotano, A.J. Performance analysis of a flat panel manufacturing line adding hybrid laser welding technology. *VII Congreso Nacional de i+d en Defensa y Seguridad. Contribución DESEID-163*. 2019.

34. Turichin, G. Hybrid Laser-Arc Welding of the High-Strength Shipbuilding Steels: Equipment and Technology. *Phys. Procedia, NOLAMP_16*. 2017.

35. Agrawal, B.P. Challenges in application of pulse current gas metal arc welding process for preparation of weld joint with superior quality. *Int. J. Eng. Res. Technol.* 2016.

36. Lancaster, J. Handbook of Structural Welding. *1st ed.; Abington Publishing: Abington, UK*. 1997.

37. Rethmeier, M. Laser-Hybrid Welding of Thick Plates up to 32 mm using a 20 kW Fibre Laser. *Q. J. Jpn. Weld. Soc.* 2009.

38. Paul Denney. Hybrid Laser Arc Welding — Has Its Time Finally Arrived? Lincoln Electric Corporation. Available on *https://www.lincolnelectric.com/de-de/support/process-and-theory/Pages/hybrid-laser-arc-welding-detail.aspx* (accessed on 30 November 2019).

39. Turichin, G.I. Influence of the Gap Width on the Geometry of the Welded Joint in Hybrid Laser-Arc Welding. *Phys. Procedia*. 2015.

40. Üstündağ, Ö. Hybrid laser-arc welding of thick-walled ferromagnetic steels with electromagnetic weld pool support. *Weld. World*. 2018.

41. Webster, S.E. Hyblas: Economical and Safe Laser Hybrid Welding of Structural Steel: Final Report. *Research Fund for Coal and Steel, Contract RFSR-CT-2003-00010; Corus UK Limited: South Yorkshire, UK*. 2009.

42. Wetzel, G. Investigations of properties in hybrid welds at 40-mm sheet thickness. *Weld. World*. 2014.

43. Bunaziv, I. The penetration efficiency of thick plate laser-arc hybrid. *Int. J. Adv. Manuf. Technol.* 2018.

44. Bunaziv, I. Laser-arc hybrid welding of thick HSLA steel. *J. Mater. Process. Technol.* 2018.

45. Wahba, M. Single pass hybrid laser-arc welding of 25 mm thick square groove butt joints. *Mater. Des.* 2016.

46. Zhang, C. Microstructure and mechanical properties of narrow gap Laser-Arc hybrid welded 40 mm thick mild steel. *Materials*. 2017.

47. Church; G. Joining Giants. Laser Systems Europe. Hybrid Welding. Issue: Summer 2016. Available online: *https://www.lasersystemseurope.com/feature/joining-giants* (accessed on 2 November 2018).

48. Verhaeghe, G. Low-Porosity Laser Welding of 12.7 mm Thickness Aerospace Aluminium; *INALCO: Tokyo, Japan*. 2007.

49. ISO International Organization for Standardization. Available online: *https://www.iso.org/standards.html* (accessed on 30 November 2019).

50. Kah, P. The influence of parameters on penetration, speed and bridging in laser hybrid welding. *Mechanika*. 2011.

51. Kah, P. The effect of the relative location of laser beam with arc in different hybrid welding processes. *Int. Congr. Appl. Lasers Electro-Opt.* 2018.

52. Bunaziv, I. Fiber laser-MIG hybrid welding of 5 mm 5083 aluminum alloy. *J. Mater. Process. Technol.* 2016.

53. Kah, P. Overview of the Exploration Status of Laser-Arc Hybrid Welding Processes. *Rev. Adv. Mater. Sci. 30*. 2012.

54. Olschok, S. Laser beam Arc Hybrid Welding of steel in thick sheet metal area. *Joining technology reports published by Aachen. Shaker Verlag*. 2008.

55. Casalino, G. Hybrid Laser Welding: A Review. *DAAAM International Scientific Book: Vienna, Austria*. 2010.

56. Olsen, F.O. Hybrid Laser-arc Welding. *1st ed.; Woodhead Publishing Limited: Oxford, UK*. 2009.

57. Üstündağ, Ö. Hybrid laser arc welding of thick high strength pipeline steels of grade X120 with adapted heat input. *Journal of Materials Processing Tech. 275*. 2020.

58. Roland, F. Advanced Joining Techniques in European Shipbuilding. *Journal of Ship Production*. 2004.

9. AUTHORS' BIOGRAPHY

C. Churiaque is an Industrial PhD student at the University of Cadiz (UCA) and Navantia S.A, S.M.E. Besides she holds the title of Manufacturing Engineer, Naval Architect, and Master in Occupational Risk Prevention and Integrated Management. She is responsible for the study of the industrial implementation of HLAW technology on butt and fillet joints of naval steel plates. Her previous experience includes research projects on FEM simulation of forming processes for aerospace industry (UCA), as well as structural calculations, stability and towing studies for shipbuilding and offshore pipeline works (DRACE Infraestructures), etc.

J.M. Sánchez-Amaya is Tenured Lecturer of Materials Science and Metallurgical Engineering at School of Engineering of UCA. At UCA he is responsible for the hybrid laser welding research line of the UCA-NAVANTIA UIC. His previous experience includes research on Laser Welding of industrial alloys.

M. Chludzinski is a Postdoctoral Researcher at the University of Cadiz (UCA). She is responsible for metallurgical and mechanical characterization of laser welded joints. Her previous experience includes development and manufacture of innovative welding equipment and methods, as well as metallurgical and mechanical assessment of metallic joints.

R.E. Dos Santos is a researcher at the University of Cadiz. Besides this he holds the title of Master of Materials Science developed in Flash Butt Welding Joint of Offshore Mooring Chains. He is responsible for FEM simulation of HLAW on butt joints of naval steel plates. His previous experience includes researches related to welding processes, development and manufacture of innovative welding equipment and methods (with patent granted), metallurgical and mechanical characterization of metallic joints.

M. Porrúa-Lara is the welding coordinator at Navantia S.A., S.M.E UN ABC. At Navantia he is responsible for the hybrid laser welding research line of the UCA-NAVANTIA UIC. His previous experience includes research on Laser Welding of industrial alloys in the naval sector.

F. Abad-Fraga works in the Shipyard 4.0 department at Navantia S.A., S.M.E. U.N. Bahía de Cádiz Shipyard. At Navantia he is responsible of the project UCA-NAVANTIA UIC, R+D+i investments, CFA (Advanced Manufacturing Center), and Naval Cluster of Cádiz in the Puerto Real and San Fernando Shipyard. His previous experience includes research on Laser Welding of industrial alloys in the naval sector.

ON BOARD ADDITIVE MANUFACTURING OF SPARE PARTS FOR THE NAVAL SECTOR

D. Moreno-Sanchez, J.D. Rodriguez, A. Dominguez-Calvo, S.I. Molina, Universidad de Cádiz, Spain
E. Corrales-Estárico, V. Casal, F. Abad, Navantia S.A., S.M.E., Spain

SUMMARY

The failure of a component of a ship sometimes requires a fast replacement to avoid delays in its operations. One of the most promising innovations to reduce the time required to replace a failed part and improve logistics is the use of 3D printers to produce spare parts onboard a vessel. Additive manufacturing technologies allow parts to be created anywhere, with almost any shape using a wide range of materials, directly from a digital 3D model. In order to create the digital model, either reverse engineering technologies can be implemented, or a database of optimized 3D components needs to be developed. Navantia, in collaboration with INNANOMAT research group, have studied in detail the change of mechanical properties of parts produced using 3D printers. This study was carried out using platforms that simulate vibrations produced in a ship, and with 3D printers installed onboard a ship. In this paper we present the process of manufacturing spare parts using 3D printers that work on a ship, evaluate the influence of sailing conditions on the quality of the manufactured parts, and review examples of spare parts produced with 3D printers onboard.

NOMENCLATURE

AM	Additive Manufacturing
FEM	Finite Element Method
FDM	Fused Deposition Modelling
FFF	Fused Filament Fabrication
ASA	Acrylonitrile Styrene Acrylate
UTS	Ultimate Tensile Strength (MPa)

1. INTRODUCTION

Additive Manufacturing technologies have become an essential part of the Industry 4.0 and are experiencing a huge growth nowadays [1], [2]. From its origins to its very first industrial uses, additive manufacturing has evolved greatly in a short time, proving that these technologies have become an important part of the fourth industrial revolution. Its development started to draw the attention of many industrial disciplines, introducing additive manufacturing in processes previously reserved to well-known manufacturing techniques and validating its enormous potential as a powerful and capable manufacturing technology, even in the most delicate and exigent applications. In this context, a great number of companies have invested in these technologies, incorporating it in their productive model [3]. This means that additive manufacturing has left the laboratory and has begun to develop as a feasible alternative to traditional methods even in a purely productive environment. Projects such the collaboration between the University of Cadiz and Navantia provide a lot of useful information, experience and knowledge to continue the research and improvement of additive manufacturing technologies in an industrial context, expanding the horizons of these technologies.

A large number of companies are starting to adopt these technologies as a very useful tool in many stages of the production scheme, not only for the conceptualization process, where additive manufacturing has been traditionally used to validate piece designs and to perform rapid prototyping, but they can also be extended through all the product's lifecycle by offering a fast alternative to produce even pieces to replace damaged components during operation or to create final parts directly [4].

Traditional manufacturing technologies are based on subtraction and conformative principles: pieces are manufactured by removing material (by carving, milling, etc.) from a raw material block or, on the other hand, using a mould. These technologies had been used for centuries, becoming an important part of the industrial processes' history through the ages. Its performance peak arrived with the development of Computer Numerical Control (CNC) machines and CAD/CAM software, integrating a tremendously efficient set of technologies. But with the arrival of Industry 4.0, traditional manufacturing techniques were not enough for certain purposes. The evolution of design brought new philosophies, which aimed to optimize the performance of the designed geometries, in order to reduce manufacturing time and material use, e.g., topology optimization or lattice structures. That meant an evolution into brand new, organic based geometries, much more complex than traditional designs. It's in this new production field where additive manufacturing is revealed to be an excellent method to overcome such difficulties. By its new philosophy, based in the addition of layers of material, additive manufacturing could set a new vision for the ways of manufacturing, and give the industry the chance to swap to a new, more flexible production model. Besides new developments of these technologies, by growing in size and capabilities to work with new, tailored materials, developed for each particular piece or process, turns additive manufacturing into a valuable ally in the new paradigms of fabrication that brought about Industry 4.0 [5].

1.1 ADDITIVE MANUFACTURING

The main principle of additive manufacturing technologies is the controlled addition or deposition of materials to conform with a three-dimensional geometry by layers.

These depositions are controlled by a numeric control software. The function of this software is to decompose the geometry into a layer structured model with certain characteristics driven by user defined parameters. This software, popularly known as slicer, provides the machines with the necessary instruction list to control the deposition of material, similar to a CNC machine tooling paths [6].

Nowadays, additive manufacturing is not limited to a single technology, but refers to a large number of different techniques. As is stated by the Society for Testing and Materials (ASTM), additive manufacturing processes can be classified in seven groups due to the technique used to deposit the material layers, highlighted in Table 1 [7]. Although its principle is the same, the deposition of materials to build a geometry by layers, the developed ways to do so, and the typology of the materials used have generated multiple and well differentiated technologies.

Table 1: Classification of Additive Manufacturing Technologies listed by ASTM.

Process	Layer manufacturing	Materials
Vat Polymerization	Photopolymer resins cured by an UV font (LCD screen, laser).	Polymers and composites.
Material Jetting	Drops of photopolymer resin cured by UV.	Waxes, polymers and composites.
Binder Jetting	Powder particles joined by a binder agent selectively deposited.	Polymers, metals, ceramics and composites.
Material Extrusion	Selective deposition of material through a nozzle.	Polymers, composites, ceramics and biological materials.
Powder Bed Fusion	Powder particles joined by thermal energy transference from a source (laser, electron arc…)	Metals, Polymers, composites and ceramics.
Sheet Lamination	Selectively cut layers of materials joined by binders.	Paper, polymer and metals.
Direct Energy Deposition	Particles or filaments joined by applying focused thermal energy (electron beam, laser)	Metals and alloys.

This differentiation, far from being an obstacle to implementation, becomes a significant advantage because this variability provides additive manufacturing with a significant flexibility, making it able to adapt itself to the manufacturing process by choosing of the most adequate technology.

1.2 THE ROLE OF TOPOLOGY OPTIMIZATION

Topology optimization has become an important ally to the process of designing geometries for additive manufacturing. This technique consists of the application of an algorithm to modify the morphology of the piece by removing a certain amount of volume or mass of the piece but maintaining or even improving its mechanical properties (resistance, modes of vibration, etc.). This is achieved by a computational study of the load distribution, given by a FEM analysis of a certain load case, using this data to determine the amount of volume to reduce. Following these steps in an iteration process, applying the same load case to the resultant reduced geometry on each step and repeating the process of volume subtraction, we can obtain a fully optimized geometry, with the minimum volume necessary to maintain the desirable characteristics [8], [9].

To perform a topology optimization analysis, the following steps are usual:

- To define a blank geometry to perform the analysis. It's necessary to carefully design a primary geometry which will be the base target of our analysis. In this geometry we define the zones in which the optimization algorithm will be applied, and the zones that will be spared in the final design. This is important to perform a design that meets the operative requirements, leaving the necessary parts untouched, such as handholds or connection parts. In this geometry we also define the zones where the boundary conditions of the FEM analysis will be applied.
- Once the blank geometry is defined, the topology optimization calculations must be configured. In this case, it's necessary to establish the part of the geometry that we want to be optimized, the part that will left out of the analysis, and the load case determined by the operative requirements of the piece.
- Following this step comes the definition of the optimization objectives. These objectives will define the conditions on which the analysis will be based and will determine the shape and volume pattern of the final piece. For this purpose, a maximum-allowed tension criterion would be applied, in case that our piece must meet some normative requirements, such as a maximum load requirement. On the other hand, a percentage of volume or mass reduction could be set, in order to reduce the amount of material to the minimum possible, if a weight or price requisite must be met.
- Next comes the calculation itself. Software analyses the geometry and the tensions produced by the loads applied, using a FEM calculation to determine the tension distribution along the geometry's surface. This information is used to determine the zones that are less affected. If these zones are situated inside the optimization zone, their associated volume will be reduced until the

limit imposed in the previous step is reached. These located reductions generate a new geometry, and the analysis is repeated until objectives are met in all the volume of the piece.

Depending on the number of objectives set, the complexity of the initial geometry or the characteristics imposed to the FEM analysis (number of mesh elements, solver type, etc.) the topology optimization analysis could need a significant number of iterations to achieve a convergent solution.

The complex, organic-shaped geometries created by topology optimization algorithms were impossible to manufacture using traditional ways such as milling or carving, due to the presence of holes, curved surfaces, and thin members. But, as remarked earlier, additive manufacturing could overcome these limitations. By using additive manufacturing technologies, it is possible to manufacture such geometries, taking advantage of the possibilities offered by the topology optimization algorithms. This forms a powerful tandem, offering the possibility of manufacturing high performance pieces with a considerable material saving. This is an advantage in certain applications where piece weight is critical, such as aeronautics and naval industry. In addition, it would help to overcome certain disadvantages produced by the anisotropic behaviour imposed on structural resistance due to its layered structure, helping to distribute tensions to reduce the critical peaks formed in abrupt geometry changes, like non-round corners or too thin members.

1.3 ADDITIVE MANUFACTURING IN THE NAVAL SECTOR

Due to its enormous advantages, the number of companies and sectors which are implementing additive manufacturing technologies is growing day by day. Sectors like aeronautics or the automotive industry have integrated additive manufacturing technologies in production models, contributing an increase in their productivity from early design stages to production itself [10], [11].

Whereas these sectors have largely implemented additive manufacturing technologies, a significant number of sectors These are in the early stages of implementation of these technologies, being immersed in an exploration and research phase in order to define the place of additive manufacturing in their production schemes. One of these is the naval sector.

Although few studies have been published using AM in Naval sector, we would point to the remarkable contribution made by Scheck et al. AM technologies are considered as disruptive technologies, already implemented in naval facilities to produce prototypes and moulds, with a great reduction in time and cost [12]. Also relevant is the development made in the Damen Shipyard Group or the research of Staino et al. setting the bases of using AM to produce custom propellers [13], [14]. Moreover, the University of Maine has recently created the world's largest 3D printed boat using a large area additive manufacturing [15].

The potential of additive manufacturing in the naval sector is notable even in the early stages of technology implementation. From early design stages to logistic support during a ship's lifecycle, additive manufacturing could be a very valuable ally to improve the processes associated with the shipbuilding and maintenance activities. Its main advantages in naval industry are:

- Possibility of offering a fast prototyping tool, providing valuable support to the design process, allowing naval architects to validate the design of multiple pieces at a lesser cost.
- AM technologies are useful for performing reverse engineering processes: it offers the possibility of analysing manufactured pieces (for example by 3D scanning technologies) and reproducing them with a fraction of the effort involved in other techniques. This is very useful to optimize and improve obsolete designs and to manufacture replacement parts in a short time, even if the original designs are not available.
- With the development of large-format additive manufacturing came the possibility of building even bigger parts, making it possible to propose the manufacturing of larger pieces of a ship, especially those with a complex geometry, where it is very difficult to do using traditional, steel-based building techniques. This could lead to a reduction in manufacturing time and costs, to easier replacements, and to the improvement in performance of the pieces produced, such as weight reductions or structural optimisation.
- The enormous flexibility of AM ensures that efficient materials could be developed to be adapted for each use, allowing the designer to adapt and create new materials in order to improve the processes and performance of the pieces produced, offering a new field of improvement.

Lines of improvement are being explored by multiple research projects undertaken by Navantia and the University of Cadiz. These projects aim to explore the integration of AM technologies in this sector and cover most of the processes associated with naval construction and maintenance. Projects such as 3DCABINS, aimed to develop designs, processes, and materials to manufacture a toilet module to be installed aboard [16], [17], or ADIBUQUE, which explores the possibilities of replacing certain ship elements, with a complex fabrication and assembly processes, with a counterpart manufactured with AM technologies. The main objective of these projects is to integrate AM in the shipbuilding process, from material development to production schemes, or even into the operational life of the ship, this being the case of the project 3DONBOARD, which aimed to develop strategies to install

AM machines onboard to print spare parts in situ. This project will be the subject of the next section, analysing the feasibility of these strategies.

2. CASE OF STUDY: ONBOARD ADDITIVE MANUFACTURING OF SPARE PARTS

One of the most important considerations in ship design is the relation of maintenance process and failure response. In every industrial process, the elaboration of protocols to reduce the impact of failures is very important to avoid the greatest number of losses as possible. A failure that means the stoppage of the production chain could cause enormous expenses in time and money and could be a risk to security. That is especially important on ships because of the greater number of elements that could cause a catastrophic failure, compromising the capacities of the ships, causing an ecological disaster, or even putting the crew˘s life at risk. For these reasons, a fast response in case of failure is essential to maintain the operational capacity of the ship. Where this is the case then AM could offer an optimal solution.

One of the most promising uses of AM in the naval sector is to use it for printing spare parts as replacements for pieces that suffered a failure or have reached the end of their service life. As previously stated, the high flexibility and low response time of AM techniques are adequate for its use as a quick response tool to reduce the impact caused by a failure, especially those which are critical enough to compromise the functionality of critical systems, causing considerable losses of time, money, or operational capacity. Shortening of time lapses between failure and repair is one of the biggest challenges in a ship˘s maintenance processes. For this, AM offers a unique possibility: the installation of machines on board in order to provide the crew with the capacity of manufacturing spare parts immediately.

The installation of AM machines on board requires:

- The fabrication of pieces in a short time, avoiding typical delays of the supply process. This is very important to ensure the operational capacity of the ship is not compromised and put its mission at risk, as well as avoiding possible long periods of inactivity of critical systems, which could mean a risk to security.
- The reduction of the amount of spare parts carried aboard the ship, reducing the need for space to store these parts and the maintenance tasks to ensure these parts remain operational.
- The increase of the lifecycle of the ship by reducing the risk of the lack of spare parts suppliers due to the technological obsolescence of the systems. This obsolescence could be a problem because of the large operational life of a ship, which could mean that a significant amount of equipment could become obsolete, making the spare parts supply very difficult, if not impossible. AM could overcome this disadvantage by the elaboration of a digital catalogue of pieces, which will be always available, and which would not be affected by a lack of supply.
- The possibility of optimizing the pieces could lead to an improvement in a ship˘s performance: weight reduction, material saving, adaption of the geometry to the operational necessities, and material developing to adapt the pieces˘ characteristics to the environment.

As we can see, there are plenty of advantages which encourage the implementation of AM machines on board, making it feasible. Possible disadvantages are:

- Operational conditions aboard a ship, especially the ship˘s behaviour at sea, could affect the mechanical characteristics of produced pieces. This will be studied in the project to determine the real impact of these factors.
- Specific training could be necessary for the operation of reverse engineering, AM machines, or the post-processing of pieces. Fortunately, AM machines tend to be more user-friendly, which would imply a reduction in the necessity of specific training.

Because of the characteristics of the marine environment and the morphology of a ship, the choice of technologies must be carefully analysed prior to the testing. The main factors that governs the technology choice are:

- Machines will be subject to a continuous state of movement produced by the response of the ship to the sea. These movements, which would be relatively violent on certain sea conditions, could be a determining factor.
- The lack of space on ships necessitates the consideration that large size installations are inappropriate as well as processes that need specific extra equipment for processing the pieces.
- Marine environments could be very aggressive to the storage the machines or materials used (high humidity, salinity, etc.).

For the reasons identified above, in this study the main technology considered is the material extrusion technology, using polymer materials, commonly known as FDM or FFF, specifically a Ultimaker S5 printer. This technology has proven its reliability in multiple conditions with a good price-quality rate, and pieces produced are ready to use, in contrast with other technologies that need long and complex post-processing operations to get the piece ready to use. In addition, the operation of FDM machines is relatively simple in comparison with other processes, and the material format (filament spools) is easier to use and storage than in other techniques.

For the validation of the chosen technology in the operational conditions in a ship, the research was conducted in two steps, firstly printing samples simulating the movement of a boat, using ASA as a material base, and in a second step, printing spare parts. The parts printed for the evaluation were standard-tensile specimens (ISO/ASTM 52921:2013) in vertical orientation (Z), the most unfavourable orientation. The results obtained in these dynamic conditions were compared with samples printed in static conditions to assess the influence of movement in a future real situation. All the tensile tests were made in a universal tensile device (Shimadzu) at 2 mm/min, following the rule mentioned.

To simulate the response of the ship's movement at sea, a specific platform was developed and manufactured in which the printer was fixed. This platform is programmed to simulate multiple vibration frequencies, a typical sinusoidal movement produced by certain wave conditions, and an intensity-variable slamming impact, which is the most violent condition a ship can face at sea, avoiding the necessity of maintaining the printer onboard and the variable conditions of the sea. Also, the same platform has sensors that measure the movement of a real sailing and is configured to allow an active compensation mode when installed in the vessel. When installed this feature will be tested in future onboard experiences.

To test the performance of the process in a real navigation environment, an onboard test was made. In this case, the experiments previously performed inland were repeated aboard a real ship to validate the results and check if the simulated results corresponded with a real situation. The experiments were performed aboard the oceanographic ship UCADIZ, the property of the University of Cadiz, with a boat velocity of 8 knots and a swell sea.

Both devices, printer and platform installed inside the vessel, are shown in Figure 1.

These tests help to validate this approach and to determine the conditions in which the fabrication could be performed without compromising the integrity of the fabricated pieces. Finally, a set of different spare parts were printed according to the demands of the shipyard and the requirements of the in-service boat. Due to the damage of the broken parts, most of them required the use of a 3D scanner and posterior virtual reconstruction. As exposed, in future, it would be interesting to dispose of a virtual catalogue with all the components of the boat, reducing significantly trials and errors and the necessity of reverse engineering knowledge by the crew.

Additionally, in some parts we use advanced design tools, such as finite elements and topological optimization, to guarantee that the replacement element will fulfil minimum mechanical specifications.

Figure 1. Onboard set of printer with compensating platform.

The results of the tests show that certain sea conditions have a negative impact on the quality of the specimens analysed, both in aesthetic and mechanical characteristics, in comparison with their counterparts printed in static conditions. This means a reduced mechanical performance of the pieces produced with dynamic movement, created by fabrication defects induced by the movement components applied to the machine during operation [18].

To corroborate the prediction of the simulated conditions new sailing tests, with a more aggressive conditions at sea, have been programmed. Besides, these new trials will allow the evaluation of the active compensation of the platform and its influence on the final result.

To create a new element it is essential to dispose of the 3D virtual file to create the printer code with the desired settings. As commented above, in the case of the 3D files of the original design of the parts not being available, it is necessary to create them by scanning and reconstructing the damaged part, or directly creating a new geometry using design tools.

When a new geometry is required, the more suitable situation is to have a part equal to the broken one. This is the case exposed in Figure 2, for a small actuator of a device, where the geometry obtained from the digitalization of a similar one has no holes and enough details to be used directly to create the code and print. These parts were printed in static conditions and with simulated vibrations, to observe the effect of the movement in the final result. Even in these small parts (around 5cm length), it is tangible that the accuracy of the element is reduced, presenting warping in its right zone, in good agreement with the tensile test results.

In other cases, when we have not access to a similar one and the damaged part have a complex geometry by the scanning, results that the digitalization cannot help complete the geometry, a reconstruction is required. The process starts with the analysis of the part, continuing with its digitalization by 3D scanning, and finishing with the new geometry reconstructed and adapted ready to be printed.

Finally, when the original parts were made with other material, usually the process will require the creation of a new geometry that satisfies the in-service requirement of the part with the new material. For this step we use advanced design tools, such as topology optimization and FEM analysis to validate the final design that also consider the manufacturing process.

3. CONCLUSIONS

Additive Manufacturing has proved to be a powerful tool to improve the shipbuilding process. Through diverse experiences, experiments and projects, the integration of AM technologies in all the stages of the naval context is nearer day by day. The advantages offered by these technologies are notorious, and prove the enormous potential of these technologies in the naval context, where several organisms such as armies are providing information about their use of the production of spare parts.

The experiments performed in this study show that is possible to manufacture by 3D printing spare parts onboard. The use of an appropriate design methodology to create these spare parts onboard, sometimes using a 3D scanner and advanced design tools, greatly help this process. However, sea conditions heavily determine the mechanical performance of the fabricated parts, reducing their mechanical properties and accuracy, due to defects caused by the movement induced in the 3D printer structure.

For this reason, we consider that a compensation of the onboard printers is required, not only to increase the final quality of the resulting parts but also to improve the lifecycle of the printers. The next step of this R&D study will be focused on the use of the compensation mode of the platform to corroborate the reduction of the movement components applied to the machine during its operation, to help to improve print quality.

This research also concludes that we need to get more knowledge about the process of onboard printing. Only a few news items have been published in the press: no extensive scientific research has been done.

4. ACKNOWLEDGEMENTS

This work was supported by NAVANTIA company (in Bahía de Cádiz Shipyard), through 3DONBOARD project ('Development and validation if spare parts made with 3D printers installed onboard of a vessel'), which was co-funded by CEIMAR Foundation. The authors acknowledge the co-funding received from Junta de Andalucía (INNANOMAT research group TEP-946, TEP-940) and European Union.

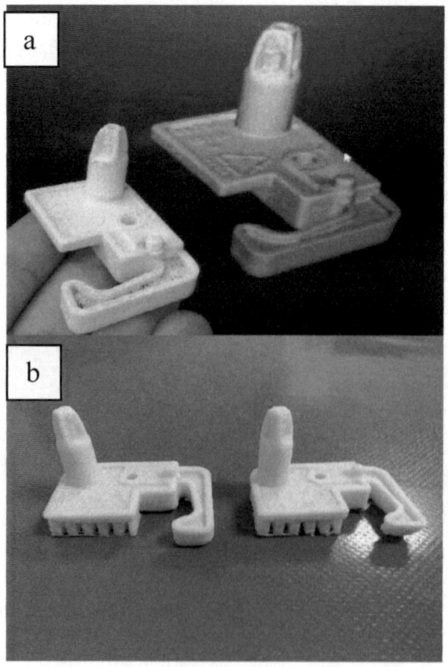

Figure 2. Element of a damage device a) original and scanned (blue image); b) printed parts without (left) and with (right) vibration.

5. REFERENCES

[1] J. (Centre for S. & E. S. L. Smit, S. (Centre for S. & E. S. L. Kreutzer, C. (Centre for S. & E. S. L. Moeller, and M. (Centre for S. & E. S. L. Carlberg, "Industry 4.0 - Study for the ITRE Committee," *Policy Dep. A*, pp. 1–94, 2016.

[2] T. Stock and G. Seliger, "Opportunities of Sustainable Manufacturing in Industry 4.0," in *Procedia CIRP*, 2016, vol. 40, pp. 536–541.

[3] R. A. D'Aveni, "The pan-industrial revolution : how new manufacturing titans will transform the world," p. 300, 2018.

[4] S. H. Khajavi, J. Partanen, and J. Holmström, "Additive manufacturing in the spare parts supply chain," *Comput. Ind.*, vol. 65, no. 1, pp. 50–63, 2014.

[5] M. Meboldt and C. Klahn, *Industrializing Additive Manufacturing - Proceedings of Additive Manufacturing in Products and Applications - AMPA2017*. 2018.

[6] J. Gardan, "Additive manufacturing technologies: State of the art and trends," *Int. J. Prod. Res.*, vol. 54, no. 10, pp. 3118–3132, 2016.

[7] ASTM-F42.91 Subcommittee, "Standard terminology for additive manufacturing technologies," *ASTM Des. F2792-12a*, vol. Vol.10.04., 2015.

[8] J. S. Jensen, "Topology optimization," *CISM Int. Cent. Mech. Sci. Courses Lect.*, vol. 540, pp. 109–159, 2012.

[9] Y. Saadlaoui, J. L. Milan, J. M. Rossi, and P.

Chabrand, "Topology optimization and additive manufacturing: Comparison of conception methods using industrial codes," *J. Manuf. Syst.*, vol. 43, no. October, pp. 178–186, 2017.

[10] L. J. Love, M. W. Noakes, B. K. Post, B. J. Rhyne, and K. T. Gaul, "Feasibility of Using Additive Manufacturing to Produce Automotive Tooling," 2018.

[11] A. Ceruti, P. Marzocca, A. Liverani, and C. Bil, "Maintenance in aeronautics in an Industry 4.0 context: The role of Augmented Reality and Additive Manufacturing," *J. Comput. Des. Eng.*, vol. 6, no. 4, pp. 516–526, 2019.

[12] C. Scheck *et al.*, "Naval Additive Manufacturing: Improving Rapid Response to the Warfighter," *Nav. Eng. J.*, vol. 128, no. 1, pp. 71–75, 2016.

[13] "World's first class approved 3D printed ship's propeller unveiledNo Title." [Online]. Available: https://www.damen.com/en/news/2017/11/worlds_first_class_approved_3d_printed_ships_propeller_unveiled. [Accessed: 20-Nov-2019].

[14] G. Staiano, A. Gloria, G. Ausanio, A. Lanzotti, C. Pensa, and M. Martorelli, "Experimental study on hydrodynamic performances of naval propellers to adopt new additive manufacturing processes," *Int. J. Interact. Des. Manuf.*, vol. 12, no. 1, pp. 1–14, 2018.

[15] "University of Maine creates the world's largest 3D printed boat." [Online]. Available: https://www.3dnatives.com/en/3d-printed-boat-university-of-maine-161020195/. [Accessed: 16-Nov-2019].

[16] D. Moreno Nieto, V. Casal López, and S. I. Molina, "Large-format polymeric pellet-based additive manufacturing for the naval industry," *Addit. Manuf.*, vol. 23, no. July, pp. 79–85, 2018.

[17] F. Javier, A. Fraga, V. Casal, D. M. Nieto, and S. I. Molina, "Fabricación Aditiva en la Construcción Naval en el," vol. 2015, pp. 1–6, 2017.

[18] V. Domínguez Calvo A., Morgado-Estévez, A., Casal López and S. Molina, "Impresión 3D de repuestos a bordo de un buque," *Actas del V Congr. Nac. I+D en Def. y Segur.*, 2019.

6. AUTHORS' BIOGRAPHY

Daniel Moreno Sánchez. Materials Engineer and Manufacturing Master. Industrial PhD student between Navantia and Universidad of Cadiz and researcher of INNANOMAT research group. He is a materials engineer specializing in design and industrial manufacturing process, mainly focused on the design of additive manufacturing, and the development and implementation of new materials for material extrusion additive manufacturing. His previous work includes design, validation by additive manufacturing, and mass production of injection plastic parts for industrial purposes.

Juan Diego Rodríguez. Naval Engineer. He is a researcher at the University of Cadiz in the INNANOMAT research group. He is part of the team responsible for designing, validating, and manufacturing spare parts and structural pieces for R&D projects in collaboration with Navantia. His previous experience includes the participation as technician in projects related to the fabrication and assembly of robotic prototypes using additive manufacturing.

Álvaro Domínguez. Design Engineer and Manufacturing Master. He is a mechanical design engineer at Navantia S.A. S.M.E. He is responsible for design parts and assemblies for mechanical and electronic equipment. His previous experience includes CAD design, machining, additive manufacturing processes, and reverse engineering at the University of Cádiz.

Sergio I. Molina. Full Professor at the University of Cadiz in the field of Materials Science and Engineering. He is the leader of the research Group INNANOMAT: "Materials and Nanotechnology for Innovation". His teaching activities are mainly focused on Materials for Product Development and Industrial Design. He has been a visiting researcher at the Universities of Cambridge, Liverpool, and Sheffield, in the UK, in the EPFL of Lausanne, in Switzerland, and a visiting professor in Oak Ridge National Laboratory, TN, USA. He is the Scientific Director of the Additive Manufacturing Division in the University of Cadiz.

E. Corrales-Estárico works in the Shipyard 4.0 department at Navantia S.A., S.M.E. She is responsible for R&D and Inventiveness. Her previous experience includes research on Additive Manufacturing in the naval sector.

V. Casal works in the Shipyard 4.0 department at Navantia S.A., S.M.E. Bahía de Cádiz Shipyard. In Navantia he is responsible for the Naval Accommodation and the R&D projects of Additive Manufacturing. His previous experience includes research on Additive Manufacturing in the naval sector.

F. Abad-Fraga works in the Shipyard 4.0 department at Navantia S.A., S.M.E. U.N. Bahía de Cádiz Shipyard. At Navantia he is responsible for the UCA-NAVANTIA UIC project, R&D, investments, CFA (Advanced Manufacturing Center), and Naval Cluster of Cádiz in the Puerto Real and San Fernando Shipyards. His previous experience includes research on Additive Manufacturing in the naval sector.

APPLICATION OF 3D TECHNOLOGIES IN SHIPS OPERATION AND TRAINING OF THEIR CREWS

M. Benito Berlinches, J. L. Muñoz Pérez, Navantia Sistemas, Spain

SUMMARY

In modern digital ships, crew members work as a team, following ship specific procedures and interacting with the machinery mainly through computer tools, with the goal of fulfilling a particular mission. Information Technology allows ship manufacturers to design digital ship management tools displaying data from different sources through homogeneous well-arranged operator interfaces, providing the user with a unified perception of the situation, enabling the decision-making process and acting as the executive means of the consequent actions. These tailored tools aim to represent the relationships between the data, enabling understanding and awareness of the situation and allowing the operator to anticipate the consequences of the available commands. Likewise, crew member training following established procedures greatly improves ship operation and equipment maintenance skills. This paper presents the use of 3D technologies to arrange ship data within a three-dimensional user interface environment, which is considered a key technique to improve and accelerate both the transformation of simple data into information and the knowledge transmission among the human teams.

1. INTRODUCTION

This paper deals with human-machine interfaces for integral ships' Information Technology management tools from the past, present and future, and specifically with the suitability of 3D representation for the development of such interfaces. The main applications under consideration will be Integrated Ship Monitoring and Control programs and their associated crew training I.T. tools, due to their global nature, the tailored development and adherence to shipyard construction schedule. To achieve this, software developers work closely with naval architects, and detail engineering and production teams. Those interfaces associated with systems or equipment to be fitted on board any vessel are not the subject of this paper, although most of the considerations are perfectly applicable.

Naval missions performance is conditioned by physical (ship and surroundings) and social (ship crew and naval base personnel) constraints resulting in social and technical systems made up by engines and people linked to structures whose combination aims to fulfil selected objectives along their existence.

Engineers that build ships and systems within ships must design with ergonomics in mind in such a way that human and machine interaction is smooth and accurate towards adequate performance considering the huge restrictions imposed by the marine environment. These ergonomic patterns are, on the one hand, defining how the users adapt to the system and, on the other hand, how users' responsibilities on each of their action or management scope are affected by the way the interfaces are implemented. For this reason, a simplistic design approach may be interesting for a specific function implementation but with the inconvenience that such a simple interface, without further chances to deduce function context, loads up the procedural needs which will require more knowledge in advance of the social structure. These procedures have to be transferred with the associated training costs on both, tools and lessons. Highly conducted contents may limit the chances to deduce consequences of the available actions and are not usually good support for system and its characteristics familiarization.

2. HISTORY

The introduction of I.T. Systems in ships' operation is relatively new considering the many centuries of naval history.

Development is continuously changing at a rate defined by the associated technology evolution in computers' processors, graphical adaptors, and visual interface devices or operator screens.

The following paragraph summarises the history of hardware elements associated with visual interface in Personal Computers.

2.1 EVOLUTION OF COMPUTER SCREENS AND THEIR ADAPTER CARDS

Computer systems have one element that poses an authentic change in the rules of the game of human machine interface. This element is the computer screen.
The first screens were introduced by IBM back in the 1970s with the invention of the single colour screen based on Cathodic Ray Tube (CRT) and image adapter cards MDA (Monochrome Display Adapter). These devices only presented characters. The beginning of the 1980s saw the arrival of the colour tube screens that had to be connected to the CGA (Colour Graphics Adapter) cards. This combination offered the software developers the chance to interface their programs through 320x200 pixels with four colours or 640x200 with two colours. In 1984 IBM introduced the EGA (Enhanced Graphics Adapter) cards providing sixteen colours with a resolution of 640x350 pixels. By the end of the same decade, screens

that were capable of rendering resolutions of up to 1024x768 pixels were well established in the market. During the 1990s graphics adapter cards increase their power with prices gradually lowering. The first adapter card with dedicated processor entered the market in 1984 and was the IBM PGC (Professional Graphics Controller) mainly developed for engineering graphics design. In 1987 IBM introduced their PS/2 line of computers with the VGA (Video Graphics Adapter) standard followed by the MCGA (Multi-Colour Graphics Array) whose main feature was that it was embedded in the computer motherboard and was not offered as a separate card. Three months after IBM˜s introduction of the VGA standard, they presented the IBM 8514 supporting a resolution of 1024x768 with 256 colours. Most widespread screens kept being CRT during the 2000s due to lower production price in comparison with innovative liquid crystal displays (LCD). LCD surpassed CRT sales in year 2008, the same year that SONY closed their last Trinitron CRT screens factory, remaining as practically the only technology available in the market. LCDs started their race in the market associated with portable equipment due to their light weight and low power demand in comparison with the CRT technology. From the first screens with passive illumination (an active illumination or reflective material back-panel is required to illuminate the LCD screen pixels) technology has evolved from the plasma displays that did not require backlighting up to the OLED (Organic Light Emitting Diode) that neither need backlight and that are slowly establishing themselves as a substitute to LCDs since they provide better colour quality (mainly in the dark range) and are thinner and lighter. Nevertheless, LCD technology still remains since disposal and recycling procedures are well established and does not suffer from ghost images resulting from long periods of fixed images in the OLED displays. Latest tendencies lead the way towards other technologies such as QD (Quantum Dot) displays.

From 1990 onwards, graphic adapters for big market were implemented embedded in the computer main board, nevertheless for special applications, graphic adapters were separated in an independent dedicated card. This is the case of computers with an intensive use of graphics or with special computational requirements such as video game computers, engineering design workstations, artificial intelligence applications or image processing for artificial vision.

Special mention is to be made of the displays of handheld devices whose quality is gradually getting better and better and which have expanded vigorously in the market entering our everyday lives and in so doing changing the way we interact with devices as well as society. These handheld devices are also ideal candidates for access to information provided by ship management systems.

2.2 ORIGIN AND EVOLUTION OF CONTROL AND TRAINING PROGRAMS IN THE NAVAL CONTEXT.

Ship platform control systems basically consist in sensors, actuators, units that connect these and that also process such signals providing some logic or PLCs (Programmable Logic Controllers) and, finally, human machine interface units. These latter could simply consist of light indicators and pushbuttons connected to the PLCs or could be as complex as a network of operator control stations based on industrial PCs showing hundreds of ship circuit synoptic views, also known as mimics, depending on the ship platform complexity.

In the process of the design and development of a ship platform integrated automation system (integrated assumes here that main operator interfaces are implemented on PC based workstations) field signals collection units or PLCs are selected at early stages from a certain brand at the convenience of the technical team. The first programmable logic controllers were introduced in the market at the end of 1968 as a substitute for relay-logic units. At that time human machine interfaces were implemented using physical panels with light indicators, gauges and pushbuttons. In 1986 a client requested that the PLCs manufacturer Beckhoff provide storage capability to his PLCs. Beckhoff found that the easiest approach was to incorporate a PC. It was early noted that the PC was able to provide more than just data storage for the control system and this was when the industrial PC era started. However, PLC data extraction into the PCs was then only possible using the tools provided by the PLC manufacturers. Later the industry standards for PLCs interfaces were created but at the beginning this interface was held captive and owned by the PLCs˘ brands. Taking advantage of this situation, and at the same time to ease the industrial applications developer˜s life, PLC manufacturers delivered their PC based human machine interfaces development suites specific to each PLC. With these development suites applications entailing different complexity in logic and graphics could be developed but always with one limitation which was that the achievable interfaces depended on the basic units of information or interaction, also called `widgets_, contained in the suite resource libraries, and then not all required functions could be implemented in a proper way.

With microprocessors˘ price decreasing, PLCs were gradually improving their performances and availability until nowadays the two types of computers coexist with both functioning in their own realm: PLCs by the equipment and PCs close to the human. Content developed for control systems using PLCs˘vendors provided tools to gradually get more and more attractive with time since HMI development environments improve their quality. Nevertheless, the 2D human machine interface paradigm remains unchanged and flat symbols and mimics are still the norm despite their intrinsic drawbacks such as counter-intuitiveness and high training demands.

Ship crews training systems are intrinsically complex because the tools supporting training activities must cover multiple skills required on board ships such as ship operation, ship maintenance, and ship damages control. That is the reason why training schools are places where ship real equipment is found together with ship compartments 1:1 scale to achieve training about damage control techniques required to address flooding or fire. Most legacy information technology tools found in training schools are those associated with ship operation and are intended to emulate those tools found in a specific real ship. It is even possible to find real software tools connected (or not) to simulation processors used to reproduce the signals from sensors and equipment into the platform control system (real or replica). Latterly 3D software techniques have entered training schools. These techniques will be dealt with later in this paper.

2.3 HISTORICAL CONSTRAINTS IN THE DEVELOPMENT OF HUMAN INTERFACE SYSTEMS FOR SHIPS.

Ship management systems, land for training or on board for platform management, have been historically determined by many factors affecting the final overall result.

For ship control systems˘ software, the first factor is the human machine interface hardware and the programming tools. The historical review provided earlier shows the determining factors that the developers have had to tackle every time a new ship construction program is envisaged. At the beginning, with operator interfaces based on ASCII characters, it was very difficult to think about images for synoptic schemes development. With the evolution of hardware and development tools, ship systems synoptic schemes or mimics started to use colours and graphical details until recently. Currently developers have access to tools and equipment capable of incorporating even highly realistic 3D images coming from many sources. The second determining factor regards ship systems which are mainly designed by engineers who have a strong technical background, derived from their work in the ship design itself. This can affect the user experience as the user will have a much lower knowledge of ship design details. The result of this is that the engineers set a very high technical entry level to the users, making it difficult for them to understand the interfaces and the underlying functionality. The effect of this is a requirement for a more complex crew training program. The system is also less intuitive and accessible. The third determining factor is the intention of the synoptic schemes, that is to say, mimics are usually developed using the minimum amount of information required to draw the `function picture_. In traditional computer systems interfaces based on flat diagrams, the information is only that which is required for that particular process. This limits the ability of the system to draw conclusions about interference or interactions among different ship systems. Let us imagine a case that a ship compartment must be isolated using remote control in order to fight a flood or fire (something that hardly ever happens but unfortunately does happen occasionally) it would be very hard to identify actions to execute in order to fight the damage just by looking into traditional flat system synoptic schemes (HVAC, electric distribution, fuel transfer and so on).

One of the determining factors affecting a ship class crews˘ training programs is the limited time allowed and the requirement for the results prior to the ship delivery for these training programs. Design strategies for development of training tools (mainly simulators) are also very diverse. Two approaches are most common. One approach is simulators based on actual equipment of the ships. This complicates the integration of the simulation rig due to the variety of ship interfaces. A second approach is simulators based on standard office PCs running the integrated platform control software for remote operators. That results in a poor familiarization with local control interfaces by the engines and equipment that do not have PC based interfaces. Training assets developed following any of both approaches try to spread knowledge about ship specific operative procedures for future crews to develop their tasks safely. Nevertheless, full awareness of physical ship layout and operation of equipment in engine rooms at local level, observing all equipment and personnel interactions, is only acquired once crews are assigned roles on board the ship and start to work on real equipment with the rest of the crew. This means an important amount of training is required after starting to work in their final posts on board with the risks this entails.

3. OPERATOR INTERFACES SUPPORTED ON 3D TECHNOLOGY

3D technology contributions to human interfaces, the subject of the present paper, are diverse in all the contexts mentioned and are listed below.

3.2 KEY CONTRIBUTIONS OF THE 3D PRESENTATION

The main contribution of presenting data over a 3D representation environment is realism. This technology applied to ship management systems eases the familiarization and knowledge of the installations and processes developed in them. In this way, interaction is more user-friendly for different user profiles such as operators, supervisors, maintainers, higher stages of command chain, students, trainers, and others.

With an IT infrastructure presenting ship data on a 3D model, equipment localization, compartments, signs, crew members, and virtually everything can be located.

The application of this strategy in the risk prevention field is key to show evacuation routes, security perimeters, and vital elements such as fire extinguishing systems or even people by means of real time location systems.

Since management systems are used extensively for many purposes and by many operator disciplines, 3D representation allows the projection of elaborated information (such as crew members flow at specific spots). Decisions can then be made about installation improvements taking into account factors such as space or accessibility (crew transit bottle necks removal for example).

For maintenance, 3D representations are the perfect reference to plan and follow up maintenance routines and are also good for remote assistances.

From the developer's perspective, 3D representations provide an unlimited multidimensional canvas for representation and realism of the models and reduces the need for using symbols and their association with ship systems that require early familiarization.

With the application of animation techniques to elements in the 3D scenes, realistic animations can be shown whose interpretation is straightforward. Different platform configurations are adequately represented, even the position in which elements and their movement are found (cranes, life rafts, doors and hatches, mooring and anchoring deck gear and more).

Operator interface now has this technique of navigation mechanisms that allow ship and systems supervision and exploration from any point of view by using camera zoom, pan, orbit and tilt movements. It also shows the operator hidden equipment parts or those parts covered by other elements such as casings and bulkheads. This would be the case for combustion chambers, interior of tanks, and other inaccessible compartments.

Overlaid information can now be shown for each scene element that can be activated by user selection, camera distance, or predefined specific events. This information can also remain in the scene depending on its relevance.
There is no doubt that attractive 3D based operator interfaces turn them into candidates to support informational or commercial activities. By enhancing realism, easy access and understanding of what is being shown, the result is a very powerful tool to reach more clients or attract inversions for new projects. Having easy access to graphical resources from the development environment enables the production of technical documents, presentations, and fliers.

Another interesting feature of this kind of technique is the opportunity to position the ship in specific natural environments such as open sea, close to coast, or mooring. This allows the extension of management system functions further outside the ship platform allowing the incorporation of other users' functions (in this case Bridge operators and navigators can be incorporated). By integrating information from bridge systems that explore the environment (maritime traffic, weather, charts, etc.) any external element can be presented in a realistic and intuitive way to any user even below decks (day-night cycles, atmospheric conditions, visibility, etc.).

Conceptual 2D synoptics are designed grouping and segregating the mimics into systems, such as cooling water system and lubricating oil. The representation of elements common to both categories such as heat exchangers can be referenced as hyperlinks that allow jumping from one system to the other, but data for each system remains separate. Ship management systems with 3D representation allow the exploration of all systems overlaid in their real locations. The convenient selective display of different systems is achieved by enabling and disabling associated layers that show or hide circuits or systems at the user's will.

3.3 HUMAN INTERFACE DEVICES FOR 3D DATA

3D representation technology allows the application of interesting and innovative human interface systems. In addition, it brings the opportunity to take the maximum out of traditional interfaces.

Traditional operator stations based on PC, one or two screens, keyboard and pointing devices, are also the developer's typical workstations, where 3D content is designed and developed. Since these stations are in fixed positions, they can easily adopt advanced graphic cards. Camera movement defining the views is configured by means of combinations of keys together with operations using the pointing device.

Apart from these traditional stations, 3D technology allows the opportunity to exploit images through more innovative devices. These include Virtual Reality immersion devices and Augmented Reality devices, thus providing a much more realistic and intimate experience with the product. These new devices have emerged from the increasing video-game industry.

- The video games market is expected to be worth over 90 billion U.S. dollars by 2020, from nearly 78.61 billion in 2017. (BestTheNews, 2016)
- There are more than 2.5 billion video gamers from all over the world. (The European Mobile Game Market, 2016)
- 80% of the total video game industry's 36 billion U.S dollars revenue in 2017 belongs to software sales. (Entertainment Software Association, NPD Group, 2017)

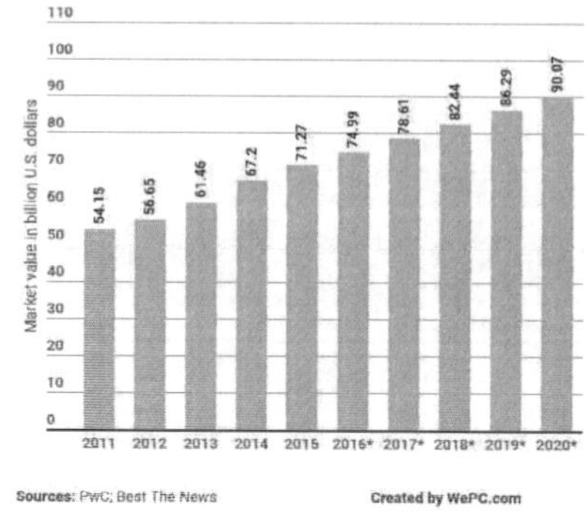

3.4 LIMITATIONS AND PROBLEMS OF 3D PRESENTATION

In this part of the paper we are going to analyse the main limitations and drawbacks that data presentation in 3D environments entail. This kind of application must run on equipment with very high graphic performances and very powerful video adaptors. Possible poor experience while navigating through the models caused by content load lags or high image refresh times is prone to get worse the more detailed the models are. State of the art associated hardware, mainly video game industry hardware, evolves in such a way that these kind of troubles are gradually less likely to be suffered by new programming techniques or by new and more powerful computer equipment.

would then be the candidate functions to be implemented using 2D techniques.

The conception and development of ship management systems in 3D imply a considerable extra effort compared with the effort required for flat synoptic systems. The economic impact on the overall project result must be minimised by the use of previously developed models (digital models from naval architects, equipment manufacturers, or external model providers). This process of re-use is complex since the model sources are not always homogenous and impose the use of software import tools and fine-tuning labour, that complicate the production process. Software tools development oriented to minimise models importing and fine-tuning processes, should be a way to minimise such extra costs.

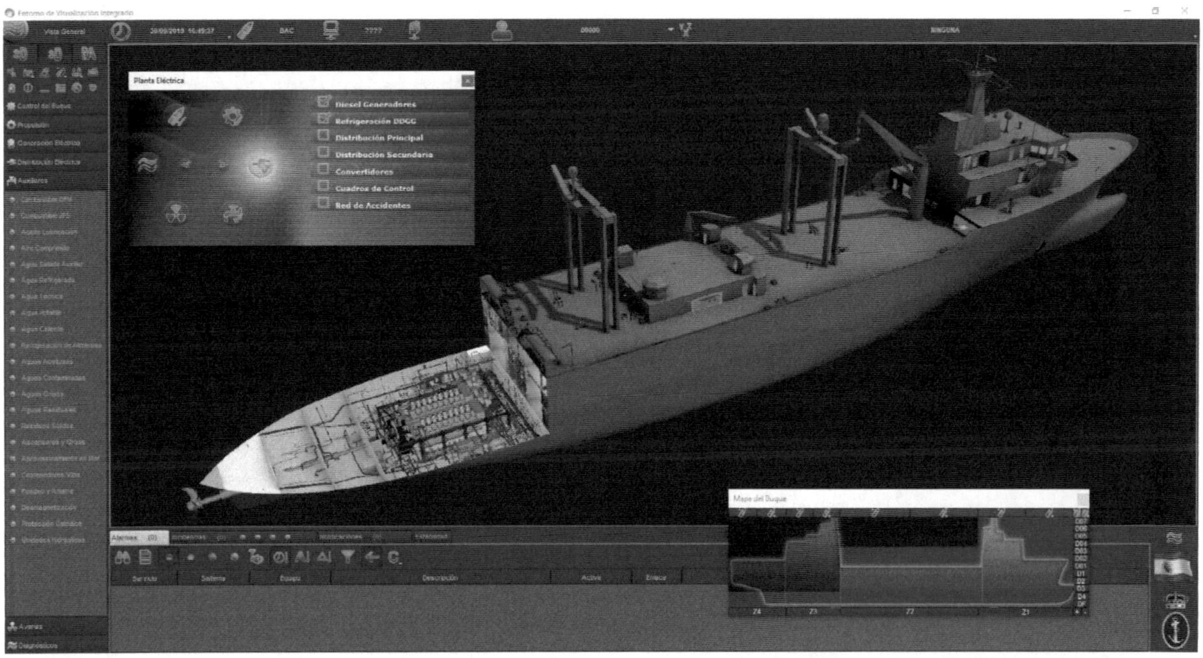

The level of detail that can be implemented is as infinite as are the details found in real ships. It is for this reason that at the moment of project conception limits to the level of detail must be established, taking into account the cost is increased considerably with higher detailed models and realistic image quality. The problem can occur in reverse. Models entering the program from other sources can easily exceed the defined level of detail of the project. A strategy must be in place in order to reduce the new item level of detail, before inserting the element into the scene and matching it overall.

Depending on what function is being implemented, it is possible that required controls cannot be inserted in the 3D scene. If that is the case, the ship management system must then rely also on flat 2D conceptual schemes and controls. Interfaces for processes involving control sequences with the intervention of more than one piece of equipment

Software development environment for 3D environments are not easy to use and stand apart from the traditional standards used to develop flat control applications. The developers˘ profile also changes and, on top of engineering skills, creative and aesthetic ones become essential to produce acceptable levels of reality and this developer profile is not easy to find in the naval engineering world.

Configuration control throughout the ship˘s life-cycle is not always maintained at the required level of detail to trigger ship management software updates on actual ship changes from refurbishments or repairs.

3D representation of the ship attracts the attention of any user: from the simple curious guest user interested in exploring ship general arrangement without any specific intention to advanced users that are aware of the ship˘s implementation details, location, interrelations, and other associated information. It is possible that in the exploration process the excess of detail can distract attention and make the user lose focus of what he was initially going to do.

5. IMPLEMENTATION CASES

This section will deal with cases of programs developed using 3D techniques. It will be possible to show positive and negatively described aspects, developments on SCADAS, and training tools for ship inside familiarisation made by the Spanish company Navantia in its control systems department.

5.1 INTEGRATED CONTROL SYSTEMS WITH 3D

Navantia˘s control systems for ships was born from the need to provide new solutions to traditional problems in the naval environment. From developing tools based on commercial development environments, Navantia left the commercial development tools to invest in their own tools to provide SCADA, implementing the Spanish Navy˘s own requirements. This was not achievable with the required quality using the market standard programming environments. It was in the year 2005 that the need arose to implement a damage reporting tool using the ship model. The damage plot in naval ships is the functionality whose intention is to draw and follow the status of every ship˘s internal incident detected and dealt with by the firefighting squads as they report to the command and control centre of its origins. This function was accomplished by drawing schemes for each ship deck, using grease erasable pencils on laminated sheets. Navantia˘s client expressed their interest about computerising the function in order to have the information accessible to every user logged in the system. At the very moment the information is available, it can be entered into the system by the damage control officer; this is how what started as a simple incidents annotation system turned into the canvas that supports any content within the ship systems. Three different views were implemented: 2D view for synoptic, 3D view for the ship model and all its circuits equipment and systems; and finally, the data view allowing the access in data query format, to the real-time information contained in the integrated platform management system real time database. All three representations are different ways of accessing the same bits of information. The result is the simultaneous coexistence of the three information presentation paradigms that complement each other by means of hyperlinks to and from any entity of those found in the ship model and covering in this way the necessity of any kind of ship operator. Damage control annotations are not made any more on static schemes but on a 3 D live model representing the status of all equipment shown in the scene. Ship familiarisation is allowed with its compartments, machinery and installation details. Elements that appear on the screen correspond with those in the real installation, achieving a quick identification of all and each one of them allowing a faster more intuitive and efficient work. One synoptic alarm associated with a certain event in a piece of equipment, has its immediate effortless direct location and visualisation in the 3D view. Thanks to the 3D view an adequate view of the data is shown without the need for previous knowledge to get the full impact of the event. Images are now self-explanatory

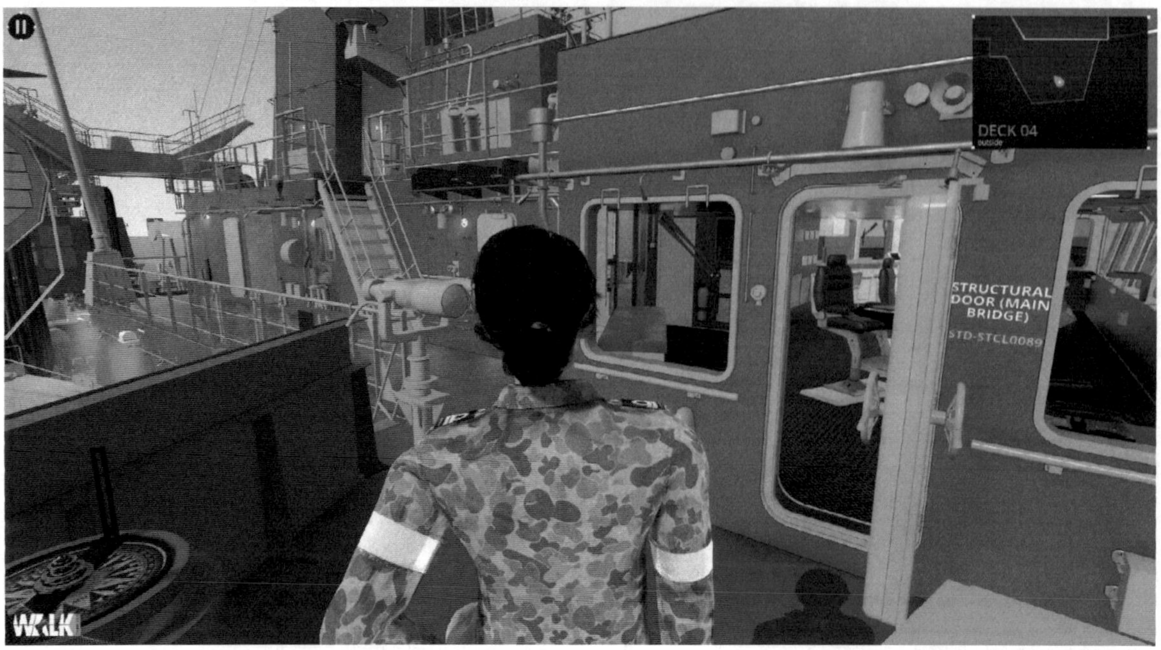

and the information within its context, not requiring previous experience. In this 3D view, optimum routes between two places in the ship are traced as they are calculated by the system considering constraints introduced by the operator. This view is also used to plot in real time the location of every crew member. This is a key characteristic in certain moments of stress such as combat stations or mooring operations where personal location and posts coverage is of essential importance. Due to the shipˇs models high level of detail this view is the ideal point of familiarisation with the physical location of systems and equipment. For this reason it is used for accessing equipment local operating panels in the On Board Training System used by operators training on how to operate the ship from the panels that stand in the engine rooms by each piece of equipment.

5.2 TRAINING TOOLS WITH 3D

Navantia provides its clients with a training solution based on simulators conceived and developed around Navantiaˇs SCADA digital platform. These training tools are integrated and take advantage of the latest status of development of associated ship project, minimising the development effort in the making of the clientsˇ training tools. In Navantiaˇs latest training projects, they make use of the power delivered by graphic engines, whose origin is in the world of video games, to re-create inside and outside aspects of their ships. This allows crews early familiarisation even before ship launch. The main objective is that students get familiar with their ship, its operation procedures and with any relevant maintenance procedures. With Navantiaˇs training system, the student is immersed inside the shipˇs virtual mock-up, containing all details that he will find in the real ship. There are several ways the student can move around within the ship. One is using a gamepad, like those used with the commercial games consoles. Another way to interact is with a keyboard and a mouse or even with a Virtual Reality equipment set. In this system, the future crew is able to walk, run, jump, crouch, turn around, and make several gestures such as hand waving, thumbs up, or say :noˇ with index finger. On exploring the inside and outside of the ship, the student gets familiar with the physical locations of essential compartments and the equipment required to execute his tasks on board. He also becomes aware of the time required to move from one place to another within the ship. On the 3D scene, the primary and secondary escape routes can be marked as can any route of interest for the student, towards the required location. Data such as functional marks and equipment descriptions or any other information considered of interest by the instructor, can be overlaid on the 3-D model. Since the models used all fall into the shipˇs digital mock up concept equipment belonging to a specific system can be highlighted, leaving the rest of the equipment with its realistic texture. Another function only possible in a virtual reality environment is the X-Ray vision that allows the user to see objects and details that otherwise would remain hidden behind bulkheads, tanks, false ceilings, engine room floors, or any other obstacle preventing the view of an item. The trainee can operate on the local operating panels that respond as in the real ship, due to the existing integration with the shipˇs physical properties and control behaviour simulator. The student can consult through the same interface, ship technical documentation associated with any equipment, manuals or multimedia contents. He can also manipulate and transport portable objects such as portable fire extinguishing equipment, hoses, portable pumps, wooden wedges to plug leaks, portable lights, and other damage control associated equipment. The studentˇs representation in the virtual scene, that is to say the studentˇs avatar, can dress in damage control associated clothing such as gloves, boots, helmets, fireproof suit, and self-contained breathing apparatus. Instructors can declare damages such as fire, smoke, gas leaks, dripping pipes, and an injured man amongst others. Inside the ship, different illumination conditions can be set, such as clear daylight, red night light and emergency illumination in case of a black ship situation. Outside the ship, different ambient weather conditions can be established: day, night, fog, rain, and different atmospheric effects. Directional sounds are emitted with distance attenuation such as the noise of engines, steps and crackling of flames and so on. The training system is multiplayer capable, since it supports several avatars in the same session which allows the training on procedures requiring teams' coordination and collaboration. Instructors can also edit scenarios and exercises which define the goals that the students must achieve and such exercises are automatically evaluated by the system. All sessions are recorded so that students and instructors can be debriefed together at a later stage. Maintenance procedures familiarisation are also possible with the implementation of such a training tool such as an engine dirty filter change, or a broken module substitution in a PLC.

6. CONCLUSIONS

Nowadays technologies open a door leading towards virtual worlds that are of real interest as a supplement of human machine interfaces in ships from a certain complexity onwards. 3D data presentation in the training field also promises to be a very promising technology to transport students inside the ships without the need to leave the training centres located on dry land. The savings in resources for large crews, such as those found in naval ships, are of no doubt. Obstacles that 3D techniques entail, although, raise certain concerns and may even pose cuts to the naval activity benefits, but are far from being insurmountable.

7. ACKNOWLEDGEMENTS

Navantia, (SEPI group) Spanish state-owned company reference in the design and construction of high technology military and civilian vessels.

8. REFERENCES

[1] "Sony to stop making old-style cathode ray tube TVs", *Wall Street Journal MarketWatch', 3 March 2008*

[2] *Polsson, Ken. "Chronology of IBM Personal Computers".* Archived *from the original on 2015-02-21. Retrieved 2015-01-28.*

[3] PC Control01|2019 "20 years of PC-based control technology - Looking back and looking forward" Hans Beckhoff interview.

[4] 2019 Video Game Industry Statistics, Trends & Data

[5] Navantia Training solutions. NAVANTIS

9. AUTHORS' BIOGRAPHY

Mario Benito Berlinches is Control Systems Engineer at Navantia, Control Systems Technological Innovation. He is jointly responsible for latest developments in Navantia's Integrated Platform Management Systems and NAVANTIS training solutions. His previous experience includes responsible of:

- Propulsion control systems for frigates.
- Prime movers and generators local operating panels.
- Frigates Power plant controls refurbishment.
- Foreign ship controls contracts management.

Juan Luis Muñoz Pérez is Ship Controls Technical Authority at Navantia. He is responsible for Navantia Control systems Technological Innovation department. His previous experience includes the following responsibilities:

- Responsible for the PC based Human Machine Interface software at Ship controls department in Navantia.
- HMI Ship Controls development for Spanish and foreign contracts such as Thai, Norwegian, Venezuelan and Australian.
- Prime movers and generators local operating panels in the F100 Spanish frigates.

NOTES ON MARITIME CYBERSECURITY IN SHIP DESIGN

C. Mascareñas and **A. I. Vázquez**, University of Cádiz. Spain

SUMMARY

European merchant traffic through the internal and external maritime lines are a critical infrastructure within the European Union, and are likely to be attacked. In this paper, an analysis of the main control systems of a ship and the possibilities of being attacked cybernetically are delineated. Conservative hierarchy in the design of the control systems is proposed and the general conclusions of the analysis are presented. The current trend towards automation implies that many ships rely on data networks and information systems. Two very important data networks are the navigation network and the propulsion control network, which can also be interconnected with each other and, in turn, to other networks. The control systems in turn depend on the SCADA equipment that has already proven vulnerable to attacks to alter its operation. Ship design strategies that can resist or mitigate the effects of a cyber attack with reduced crews are discussed.

1. INTRODUCTION

International trade and freight transport have acquired great importance with the phenomenon of globalization and the integration of economic blocs.

Maritime transport is one of the oldest ways by which goods have been transported from one place to another, considering it is one of the most used methods due to its great capacity to cargoes transport.

Maritime transport is done through different types of vessels according to the type of merchandise that you wish to reach your destination. Ships can be classified according to their speed, load capacity, dimensions and how they are structured to store the products that will be transported in them.

But all ships have in common the need for propulsion and the rudder control.

2. INTRODUCTION TO THE MARITIME CYBER SECURITY PROBLEM

Ship navigation and control systems are increasingly dependent on mission-critical information systems and industrial control systems.

In 2013, the Spanish government approved the national maritime cybersecurity strategy. This strategy indicates the following, in its line of action number 5, on the improvement of cybersecurity in the maritime field:

"The essential nature of ICT in the maritime field requires that concrete actions be established within the framework of cybersecurity to contribute to the improvement of national maritime safety standards. It is necessary to foster a comprehensive approach to cybersecurity based on the assessment of specific cyber risks and threats in the maritime field, as well as the identification of all critical assets in the sector".

The current trend towards automation implies that many ships rely on data networks and information systems for the following functions [1]:

• Navigation: ECDIS, AIS, Radar.
• Command and Control.
• Engine: propulsion, fire protection, lighting, bilge pumps, alarms.
• Cargo: cargo location control, ballast tanks, ship stability.
• Internal communications and telecommunications: satellite, VHF/HF/MF radio.
• Management systems: ERP, crew, fuels, accounting.
• Maintenance and inventory systems
• Social and entertainment networks.

The maritime sector is critical in Europe. In 2010, 52% of freight traffic in Europe was carried out in the maritime sector [2]. Furthermore, the maritime sector is increasingly dependent on information and communications systems (ICT) for operations such as cargo management in ports, vessels traffic systems (VTS), and the navigation and propulsion on ships [3].

In November 2011, the European Information Security Agency (ENISA) published an analysis report on cybersecurity aspects in the maritime sector. The main conclusions of the report were: were:

1. The level of knowledge and concern in cybersecurity in the maritime sector is very low or non-existent.
2. International regulations from the International Maritime Organization (IMO) only consider aspects of physical maritime safety. Therefore, consideration of cybersecurity measures is also recommended.
3. It is recommended to perform a risk analysis of information and communication systems, ICT, relevant to the marine sector.

The Transportation Sector Systems Plan defines cybersecurity as "the prevention of damage, unauthorized use and restoration of information and communication systems to guarantee their confidentiality, integrity and availability" [4] [5].

In July 2015, the United State Coast Guard (USCG) published its cybersecurity strategy which identifies three fundamental priorities for the defence of the maritime domain [5] [6]: Defend Cyberspace, Enable Operations, and Protect Critical Infrastructures.

The guidelines to prevent deliberate attacks on ships and port facilities are defined in the “International Ship and Port facility Security code, ISPS” regulations approved by IMO in 2002 [7] [8].

The threats considered in the ISPS code are mainly physical [7], since in 2002 the broadband networks were not as widespread, and embedded local area networks or integrated navigation systems were less prevalent.

In addition, the regulations for the safety of ship operations are defined in the "International Safety Management Code, ISM" [9] [10] code initially approved by IMO in 1993 and mandatory since 1998.

Here, we have to clarify that, in the maritime environment, the Spanish word “Seguridad” has two meanings in English: Security and Protection. This differentiation is very useful to undo ambiguity when speaking and writing about rules or laws applicable to the boat.

At sea, whenever the word Safety has been used, we have referred to the Protection against risks to the safety of human life, of the loads transported by the ship or for the health of the environment. It is worth remembering that the International Maritime Organization was formed in 1912 after the sinking of the Titanic passenger ship.

However, the meaning of Security, or protection against human threats, does not arise until the emergence of organized crime and modern piracy, for example, with the hijacking of the Achille Lauro cruise ship in 1985 and internationally since the Atalanta operation in 2008 [11].

3. STATE OF THE MATTER

Let's make a small review to the ship's critical control systems.

3.1 NAVIGATION SYSTEM

With the excessive integration of sensor systems with actuator systems on a ship, thanks to automatic control systems, there have been two negative effects: on the one hand, the reduction of crews and, on the other hand, the reduction of federated equipment (Figure 1) which share the information generated by them. In today's fully integrated systems, multipurpose and reconfigurable screens appear (Figure 2) that adapt to the operator's taste and custom, regardless of where the data to be presented comes from. Obviously, these screens are governed by computers with software adapted to the function they must perform, and are susceptible to a caused or accidental (unwanted) cyber attack.

We are talking about fundamental equipment such as the log (boat speed meter), the sounder (keel-to-bottom distance meter), the gyroscope (true course sensor) and the autopilot (electronic helmsman) or the ECDIS (electronic chart display). Auxiliary equipment, such as radars (obstacle detectors), GPS (global position sensor), AIS (automatic ship identification system) or satellite communications and ship radio communications systems.

All share information whose most important is the position of the ship that is obtained by GNSS.

The position of the ship can be integrated with electronic cartography (ECDIS) to show the position of the ship on the screen, instead of drawing it with pencil on the paper chart, and the ECDIS program allows you to program safety sectors, around the ship. This safety measure was done on the radar screen manually, in the past.

If the ship approaches the coast, or a shallow water zone, the ECDIS can set off a programmed software alarm, so that the Sounder is automatically activated to confirm the distance to the bottom under the keel, given that the reflected measures in the Nautical charts are low tide data.

Therefore, we already have two pieces of equipment that must be essential protected, the GPS receiver and the ECDIS Control Program and its nautical charts data bases, since any interference to the first set or the data bases corruption of the second can cause the loss of the ship by grounding or, at least, a large economic and several environmental losses.

The network described above is the navigation network but this, in turn, can be integrated with the engine control network or propulsion control network (Figure 3).

3.2 PROPULSION SYSTEM

The purpose of the propulsion control network is to keep the ship's engine or engines running at the speed necessary for the boat to reach and maintain the required speed. Today, the propulsion depends on the Integrated Navigation System to deliver the cargo at the appropriate time and place, regardless of the boat's variable hydrodynamic conditions or the weather and the changing state of the sea.

Together with the ship's steering system, the propulsion control system is indispensable for a merchant ship to complete its mission and deliver its cargo 'when and where' it has been ordered.

But marine propulsion systems can be diesel, LNG, steam or electric, and even a mixture of them, so they must have a series of auxiliary equipment, such as water treatment plants, pressure pumps, electric generators, heat exchangers, boilers, shaft generators, static frequency converters, etc, which must be included within the Propulsion Control System [12].

In general, all ship machine devices have duplicate or even triplicate controls, called "local" when the manual control is very close of its equipment, "remote" when, with manual operation, they are in the engine room or control room, grouped with those of other pieces of equipment for reasons of ergonomics, "automatic" when they are remotely controlled from computers installed in the engine control room, but with manual operation, and "integrated" when there is an activated reaction by a programmed signal and this program is running from a computer located in a different local area network.

For safety, the remote control and local control must always prevail, giving rise to the 'emergency system' of the propulsion control, in the event of a cyber attack against the integrated propulsion control system, but, of course, the cyber-attack must be detected. before irreparable damage occurs.

3.3 STEERING SYSTEM

Previously, we mentioned the Vessel Steering System, which is responsible for placing the rudder blade at the right angle, constantly, in order to the ship reaches the destination and complete its mission [13] [14].

At the beginning of the twentieth century they were divided, for vessel's safety, into Manual, if the wheel was handled by hand, Automatic, if the Autopilot was the one who decided where to put the rudder (Figure 4), and Emergency, if the sailor acted on the rudder locally. But with Integrated Navigation Systems the autopilot is not physical device, it may be a program and this program may be corrupted by a virus or malware.

The greatest danger of cyber-attack lies in an erroneous course (different from the orderly heading) not calculated by the Pilot or guard officer but by the ECDIS, instead, if the autopilot is a physical electronic device and not a “software set”, the electronic autopilot could not be corrupted.

The gyrocompass is so important. A bad detection of the True North or the change of course detection error will cause a wrong rudder response and the realization of an unwanted manoeuvre with the collision or grounding possibility.

While the gyrocompass is an electromechanical sensor element that responds to natural phenomena [14], the data transmitted by it are attainable through local area networks based on NMEA0183, NMEA2000 or RS422, as well as they can be erroneous if they receive latitude and speed data from other computers that are being attacked in turn.

3.4 GMDSS RADIO

Finally, we have the ship's external communications systems, which we will divide into GMDSS and Non-GMDSS.

GMDSS systems or mandatory installation equipment under the Convention regulating the Global Maritime Distress and Safety System [15] [16] are conventional analogue maritime radio communications equipment equipped with a digital selective call controller- decoder (DSC) and a satellite communications transceiver that only transmits data at low speed through the store & forward system, through coastal earth stations of the Inmarsat system.

They are not very target sets, because they only integrate with the GPS receiver to provide the position to be introduced in the DSC in case of a Distress, but they can receive fake messages transmitted by a delinquent station, such as orders to change the load/download harbour, fake cargo or distress messages to delay cargo delivery or access to a convoy on date and time.

Non-GMDSS systems are usually satellite communications equipment that can be integrated with the Integrated Navigation System and allow high-speed digital data and voice transmission, sometimes being used to rectify the ECDIS program or to update databases and firmware of other equipment. They can be real attacked to cause economic losses to the ship owner.

These devices (LRIT, Long-range identification and tracking of ships) can automatically transmit messages, such as the position of the ship, its speed, and destination port, to provide vessel tracking information to the Rescue Societies when they cross large seas and oceans. They can also send this, or other, information to unauthorized users if they are infected with malware.

The UAIS (Universal Automatic Identification System) [15] is used to broadcast data as the ship's name, heading, speed, cargo, and destination across short ranges to nearby ships (twenty nautical miles), allowing this to be presented on the ECDIS screen and for the performance of automatic naval kinematics calculations to determine the CPA (Closest Point to Approach) and TCPA (Time to CPA) to activate the programmed alarms mentioned above or make course and/or speed changes manually or automatically.

3.5 VIOLATION OF MARITIME SAFETY DATA

The option of tracking merchant ships has become very popular among the owners or brokers, since knowing the exact location of the vessel that moves their products is a great source of information for both the charterer and their customers. See Figure 5.

The applications most used to locate boats are:

Vessel Finder: It is a free tracker for ships with AIS systems. The website www.vesselfinder.com has a very friendly and simple interface. Simply input the name of the boat and it will give you a series of data response. If you have more specific data, such as the MMSI or IMO number, the application will quickly show you the ship's location, trajectory and basic data [16].

Marine Traffic: This is a page for the use of tracking ships with AIS system. The www.marinetraffic.com page in its Spanish version is one of the best pages to locate boats. It has a live map and after performing a search by name and choosing an option the page shows complete and detailed information of the vessel to be tracked [17].

Localizatodo.com: is similar to Marine Traffic, including aircraft, amateur radio, meteorological aids and land vehicles [18].

OnCourse: boating & sailing nav (mAIS) is the other application designed by Marine Traffic. It displays a log of maritime traffic around a vessel. It is ideal for planning and sharing and interacting with other sailors [19].

These are the great applications available to track ships, very useful applications for everyone who owns a boat or is an enthusiast of them, but they are also useful for the criminal or cybercriminal.

The first question that we should ask ourselves is: How is this type of pages allowed when the Radio Regulations of the International Telecommunication Union prohibits the interception of radio signals and the use of information not dedicated to the general public?

The use of ship positions, their names, digital selective calling codes, departure and arrival ports, the cargo they transport, speed, and heading are data that are not transmitted for use by the general public, but on the occasion of maritime and environmental safety. Who has given the interceptors permission to intercept and inject the information of their vessel on the Internet?

4. PROPOSED SOLUTIONS

In the twenty-first century, in the decade where network technology is reaching its highest level of development and where research on the Internet of Things (IoT) is encouraged, as well as unmanned vehicles and the expectation of Unmanned cargo ships, we can say that a ship can be delayed, hijacked, stranded, or sunk, as result the excessive automation integration that can produce the losses in human lives, the ship and its cargo, and irreparable damage to the environment. We propose a conservative setback for the technique but we are not positioned in an anti-automation. The risk exists and is there.

The solutions that we propose are simple and can coexist with total integration. They consist of the specific technological training of the crews and the increase of the control in the automation of the ship, with the possibility of isolation between integrated systems and also of the equipment on the command and control system of each one, that is, the obligation to have screens and physical controls for all of the equipment that can be attacked and whose value is essential for safe navigation.

For the first case, the specific technological training of the crews, we propose the return of the Radioelectronic Officer on board, as a true ICT specialist, since it dominates the technology necessary to prevent and attempt to solve electromagnetic and computer attacks.

Since the implementation of the Degree in Radioelectronic Engineering at the University of Cádiz, the members of this Research Group have placed great emphasis on the teaching of spectrum analysis techniques, EMC [20] and defence against-Jamming techniques or the use of SDR receivers (software defined radio) [21] [22] [23] [24] with nautical application.

On the other hand, when teaching the subjects of Maritime Telematics and Internal Communications, the cadets are immersed in the routing techniques and local networks of the ship, not to mention the typical of Radio Engineering, Radio Electric Systems for Navigation Assistance, Electronics, Ship Steering Systems, GMDSS Equipment, Radio Communication Practices, Electronics, Maintenance, Regulation and Control Systems, etc. [25].

Regarding the hierarchy of control systems, the approach is shown in Figure 5, where we want to indicate that the Operator's Knowledge and Competence are inversely proportional to the degree of risk of the system being attacked under his supervision.

When the knowledge of the parameters and the design of the Integrated System is left to the Design Engineer of an external company, the degree of knowledge of the operator people does not necessarily have to be broad. This makes possible the reduction of skilled crews.

It should be envisaged that, in the case of cyberattack, the systems could be unlinked, either by automatic detection systems, or manually by pressing a button, trying to isolate the threat in a single infected system. Then it would be necessary the use of the ship's internal communications systems to transmit orders orally.

Even so, in the case of the Navigation System, it is worth considering the use of a secondary navigation bridge with physical pieces of equipment interconnected with each other, exclusively for the transmission of data, as seen in Figure 1, and that could be used immediately in the event that the Integrated Navigation System falls under a real attack.

In this Secondary Bridge, the nautical charts will be made of paper and must be manually corrected, as well as federated GPS receivers, sextants and log watches as an alternative. As you can see, the level of technical preparation and competence of personnel begins to increase.

In the case of Engine personnel, who they have always been more conservative than other sailors, it is necessary to maintain the Engine Control room and to be able to completely isolate it from the Machinery Control installed on the Bridge in case of attack. It must be possible the total federation of the automation of the propulsion plant and the control of auxiliary systems, so that it is not necessary to move to the Remote and Emergency systems, since the reduction of Officers and their auxiliaries (Oilers) is such that they would make their work very difficult until they reach port, even counting on an Electro Technical Officer [26], who would not be prepared enough to try to alleviate a cyberattack.

Panés [27] states 'It is when the ship enters into operation, that cyber-attacks will most likely occur (although they could also occur during the construction stage), and he follows These attacks will come in the form to disrupt the three properties of the security of the information in computers: confidentially of the information on board; integrity of the data which is stored or used on board; or availability of the systems on board'.

Any measure taken on the security of the ship will increase the safety of the ship, its cargo and crew.

5. CONCLUSIONS

The effects produced by a Security attack on a ship is of such importance that it puts the Safety of the ship and the Environment at risk, not only causing an economic or ecological loss, but also the loss of human life.

When a company or entity on land is threatened, the external services of other companies and computer devices that prevent or restrict the attack can be contacted. But not on the ships. It is the same crew who must try to minimize disasters and maintain control over all systems.

The interest about the possibility of cyberattacks and loss of the control of one or more of the integrated systems referred to in this document, must include not only to the crews of each ship, but also its ship owners, shipyards, control systems engineers, training centres, and state and private inspection centres, as well as all the personnel who has to do with the maritime world.

6. ACKNOWLEDGEMENTS

The authors would like to thank the PAIDI TIC191 "Signals, Systems and Naval Communications" Research Group, for their financial contribution which enabled this and other lines of research to be pursued by both professors.

And they would like to thank, also, RINA for the acceptance of this paper that could be rather marginal to the theme of Congress.

7. REFERENCES

1 Internet reachable at http://erawat.es/es/ciberseguridad-maritima-en-espana

2 Internet reachable at http://nornav.custompublish.com/e-navigation-workshopconference.5165191.html

3 http://erawat.es/es/ciberseguridad-maritima-en-europa

4 Internet reachable at USCG ALCOAST 323/13: http://www.uscg.mil/announcements/alcoast/323-13_alcoast2.txt

5 Internet reachable at http://erawat.es/es/el-marco-de-ciberseguridad-y-el-sistema-de-transporte-maritimo

6 Internet reachable at USCG cyber strategy: http://www.uscg.mil/seniorleadership/DOCS/cyber.pdf

7 Internet reachable at http://erawat.es/es/como-incorporar-la-ciberseguridad-maritima-en-pbip-e-igs

8 Internet reachable at https://www.boe.es/buscar/doc.php?id=BOE-A-2004-15290

9 Internet reachable at http://www.fomento.es/NR/rdonlyres/FD0AEFA6-E601-4470-B6EE-A41E8393DDEC/121763/CODIGOIGS.pdf

10 Internet reachable at https://www.boe.es/diario_boe/txt.php?id=BOE-A-2015-5852

11 Internet reachable at http://www.emad.mde.es/MOPS/060-Yibuti-ATALANTA/

12 Internet reachable at http://articles.maritimepropulsion.com/article/First-Aircraft-Carrier-Motor-for-the-Queen-1844.aspx

13 MASCAREÑAS, C. 'Sistemas de Gobierno del Buque'. *Servicio de Publicaciones de la Universidad de Cádiz.* 1998.

14 MASCAREÑAS, C. 'Aportaciones a la Identificación de la Dinámica de un buque y sus sistemas de gobierno utilizando un simulador de navegación'. Tesis Doctoral. *Servicio de Publicaciones de la Universidad de Cádiz.* Formato electrónico. 1998.

15 Internet reachable at http://www.fomento.es/NR/rdonlyres/A59AE233-E204-4C62-A85A-172F038C156D/23109/RD_1185_2006.pdf

16 https://www.vesselfinder.com/es

17 https://www.marinetraffic.com/es/p/%3C

18 https://www.localizatodo.com/html5/

19 Internet reachable at https://play.google.com/store/apps/details?id=com.marinetraffic.iais&hl=es

20 MASCAREÑAS. C. 'Manual Básico de Sistemas de Comunicaciones Marítimas'. 2ª Ed. *Servicio de Publicaciones de la Universidad de Cádiz.* 2011.

21 Internet reachable at http://marconi.uca.es/IDT15-BAKKALI.pdf

22 Internet reachable at http://marconi.uca.es/CISTI-MARCONI-DEFINITIVO-1.pdf

23 Internet reachable at http://marconi.uca.es/jiduca%20mascare%C3%B1as%20comunicacion%20oral.pdf

24 Internet reachable at http://marconi.uca.es/demostrador%20vuhf-marin-definitivo.pdf

25 Internet reachable at http://www.uca.es/wp-content/uploads/2017/05/PE-Ingenieria-Radioelectronica.pdf

26 Internet reachable at https://www.boe.es/boe/dias/2012/03/19/pdfs/BOE-A-2012-3857.pdf

27 PANÉS, R. 'Maritime Cybersafety: another approach to cybersecurity in ships with cyber-enabled systems'. Singapore. August. 2019.

8. AUTHORS' BIOGRAPHY

Carlos Mascareñas is Assistant Professor at the University of Cádiz. Department of Navigation Sciences and Techniques and Shipbuilding. On the Radioelectronic Engineering undergraduate degree program, he is responsible of the training of the Radio Electronic cadets for the Radio Rules, GMDSS and Radio-communication Practices subjects. On the Marine Transport masters program, he is responsible for Projects, Communications Systems, and Telematics subjects. His previous experience includes his job as 1st Class Radioelectronic Officer, and Combat System Engineer, and he researches into radiation, electromagnetic compatibility, and Software Radio Defined (SDR) sets integration.

Ana I Vázquez is Substitute Professor at the University of Cádiz. Department of Navigation Sciences and Techniques and Shipbuilding. On the Radioelectronic Engineering undergraduate degree program, she is responsible of the training of the Radio Electronic cadets for the Radio Engineering and Steering Systems subjects. On the Marine Transport postgraduate program, she is responsible for Signal Analysis courses. Her previous experience research into radiation and electromagnetic compatibility. She is now writing her doctoral thesis about Remotely Operated Vehicles.

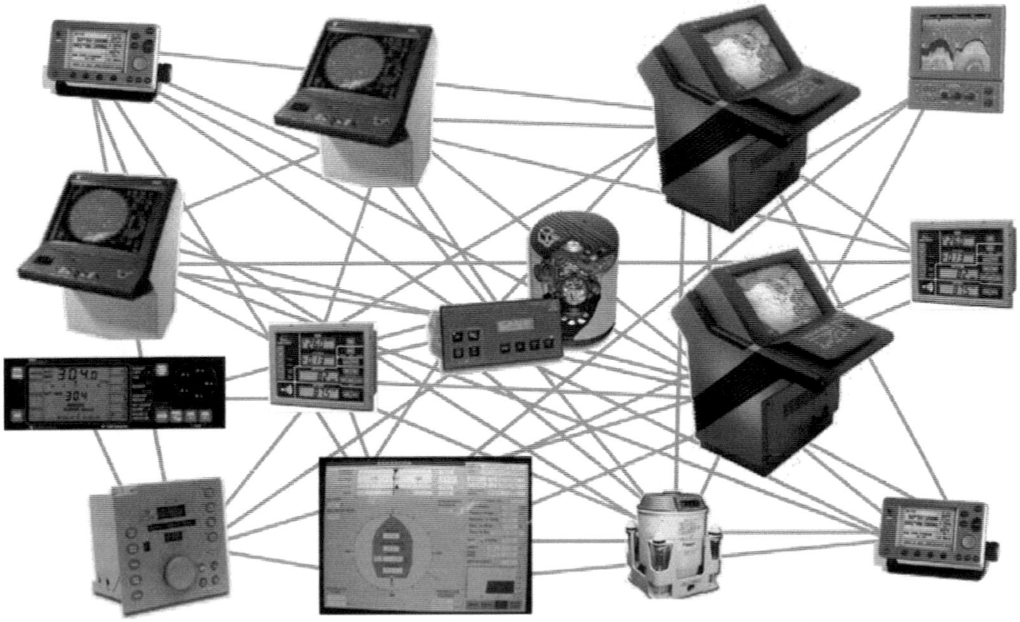

Nr.1 Integrated Navigation System with non-integrated sets (only interconnected).
Courtesy of Det Norske Veritas.

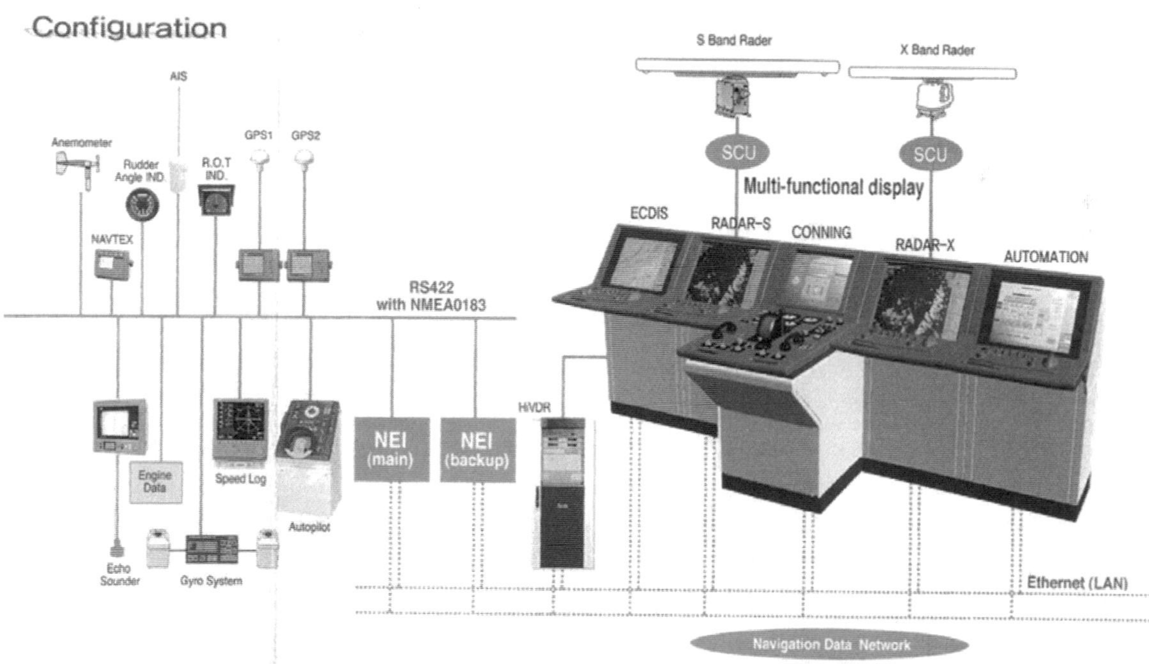

Figure Nr.2 Simplified block diagram of an Integrated Navigation System.
Courtesy of Det Norske Veritas.

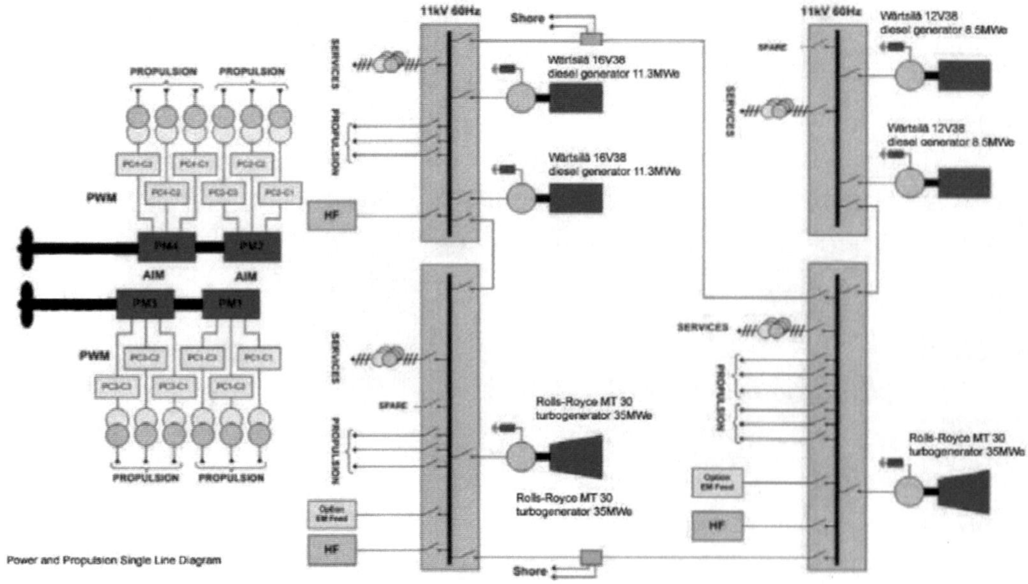

Figure Nr.3 Simplified block diagram of a Propulsion System.

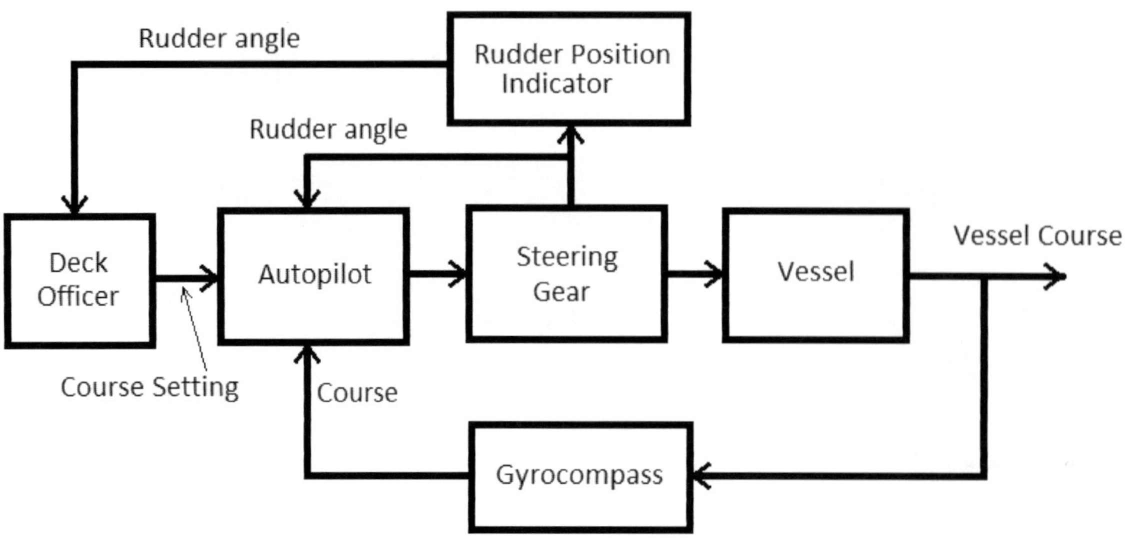

Figure Nr.4 Simplified block diagram of an Automatic Steering System.

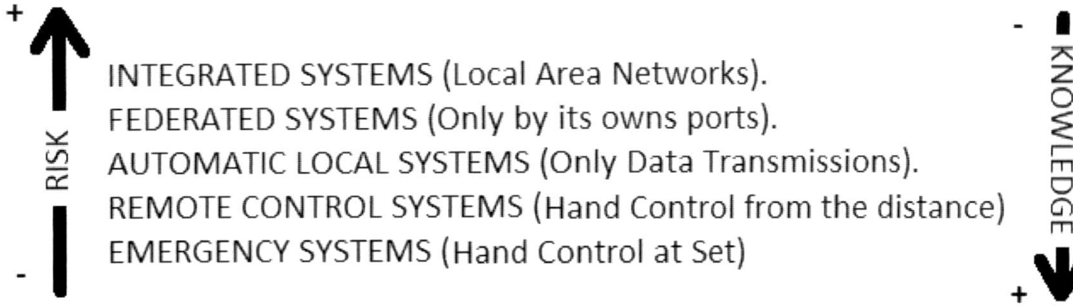

Figure Nr.5 Automation vs. Crew Knowledge.

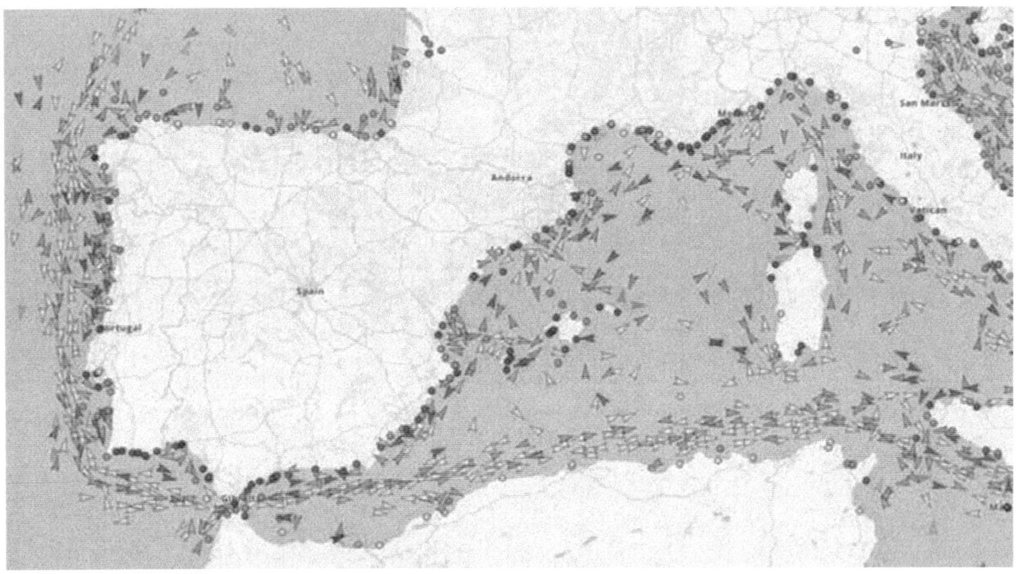

Figure Nr.6 Screen of a vessel tracker software.

ANALYSIS OF THE FLUVIAL TRAFFIC OF CRUISE SHIPS FROM AIS DATA: MOST CROWDED RIVER BASINS AND WASTES PRODUCTIONS

I. Vicente-Cera, A. Acevedo-Merino, J. A. López-Ramírez, E. Nebot, INMAR-CEIMAR, Spain

SUMMARY

Cruise tourism has experienced a significant growth in recent years, both in the size of its ships and in the number of transported passengers. Although it is a well-known market in terms of its social impact, it is still largely unknown scientifically in terms of its environmental impact. There are some studies that address the challenges of maritime traffic of these vessels but very few examine vessels' navigation in rivers. From statistical analysis and big-data methodologies developed for the marine environment, new methodologies have been adapted and developed from the combination of AIS (Automatic Identification System) and GIS (Geographic Information Systems) for the study of the fluvial environment. This paper reports on the state of the traffic of this industry at a global level, as well as its economic, social, and environmental impacts. It has been found that the ten basins most navigated accumulate 95% of the navigation hours of the entire fleet, all of them being in Europe. It has been found that the 296 river vessels studied produce 302kt per year of CO_2, 3hm^3 per year of wastewater, and 113kt per year of waste. The river cruise sector, which transports barely 10% of cruise users with respect to the ocean sector, accumulates around 20% of its global economic impact, €8,955 million per year.

NOMENCLATURE

AIS	Automatic Identification System
GIS	Geographic Information System
GRT	Gross Register Tonnage

1. INTRODUCTION

International transport is a key global economic sector from which major social and economic benefits are directly derived, such as the delivery of a wide range of goods and services. It also provides employment, mobility, and leisure opportunities for both commercial and personal purposes [1]. The cruise industry is part of this great system as one of the fastest growing in recent years, offering a complete package of mobility, leisure, and accommodation, combining the strength of the three sectors but creating a unique environmental concern.

In the last two decades the number of people who have chosen to spend their holidays on board one of these ships has grown exponentially, with at least five new ships joining the service each year. In 2014 the capacity of active cruises amounted to 453,000 passengers, a figure that by 2015 had grown by 19,000 new seats. The cruise industry estimates that by the end of 2018 the total capacity of the fleet was 537,000 passengers. With a total value currently estimated at €5 10^{10} million with 26 million unique passengers it is anticipated that there will be an estimated 27.6 million passengers in 2020 [2].

The largest ship built to date is the Symphony of the Seas, capable of carrying 8,880 people (6,780 passengers and 2,100 crew members) at the same time. On its eighteen decks passengers can enjoy swimming pools, theatres, cinemas, restaurants, shops, accommodations, etc., which makes them true floating cities, not only because of the large number of people they can transport at the same time but also because of the enormous amount of services offered in proportion to the space available.

But not all ships correspond to the description above. In addition to these giant ships there is a variety of them in all sizes and shapes, which move around the known world on seas, oceans, and rivers. This type of tourism chooses the tourist destinations that are most attractive to its passengers, in addition to connecting cities of special cultural charm, entertainment, and commercial interest. Unfortunately, most of the destinations chosen are those that, in turn, are most sensitive to economic, cultural, and environmental disturbance. The islands of the Caribbean Sea, the waters of Alaska, the Mediterranean Sea, the Nordic Fjords, the coasts of many small islands, or the banks of navigable rivers around the world all have the constant presence of these ships, and the environmental pressure that accompanies them.

River cruise traffic is often forgotten when talking about the cruise industry and is especially unknown in terms of the global understanding of its patterns and behaviours. Although the cruise market is well-known [3, 4], there are very few scientific investigations of this topic. There are a large number of navigable rivers in the world on which a multitude of river cruises take place. Since ancient times these rivers have been a valuable resource for the transport of passengers and goods [5]. It is impossible to forget how important rivers have been for human development in the cradles of civilization: Tigris and Euphrates (Mesopotamia), Nile (Egypt), Amazon (Brazil), Guadalquivir (Spain), and a long *et cetera*.

Beyond the landmarks of antiquity in which rivers and canals were really relevant for the development of nations, their adaptation as a means of modern transport began in the twentieth century. In this century the advances in hydraulic engineering allowed the modification of great portions of rivers to make them more navigable. After the commercial boom of this epoch in the centre of Europe the use of rivers like the Scheldt, Meuse, or Rhine meant that 85% of the transport of both people and goods was fluvial. During the sixteenth to eighteenth centuries improvements in river engineering, combined with a growing need to shorten travel times between nations, led to the resurgence of channelling technology not only to make navigable river regions but also to artificially channel previously unconnected areas. A relevant example is the case of the Briare Canal that connects the Loire and Seine rivers along 57 km [6]. Until the nineteenth century river transport did not stop growing, reaching unprecedented numbers of navigable portions of rivers: France (8,500 km), Germany (7,339 km), and Finland (6,120 km) [7]. In the United States of America this development took place later and by the end of the same century it only had about 6,000 km of navigable inland waters. An example of this is that today they still have only two large navigable basins (the Columbia and the Mississippi).

As with the oceanic industry the great inventions of the industrial revolution changed the world panorama drastically. The development of the railroad and the steam engine caused the decline of this form of transport, as they were faster, time being a key and constant element in this industry. It was not until the course of the two great wars that river transport regained strength, due to political instability, the risks of navigation by sea routes, and the need to transport large quantities of supplies to remote regions [8]. Cruise ship tourism, as we know it today, emerged from the need to offer new experiences in a booming tourism sector: enjoy the cultural charms of the past and otherwise inaccessible natural landscapes.

The significant increase in cruise ship activity and the associated potential environmental impacts make it necessary to carry out an in-depth analysis of the worldwide traffic of this type of vessel, from a scientific and technical point of view, and to measure the environmental impacts of both its intense traffic and its stay in port. It is important to know the generation of wastewater, solid waste, discharges, fuel and energy consumption, and gaseous emissions, etc.

Although in the maritime environment this analysis has already been resolved both for the study of maritime traffic and its associated environmental impacts [9–11], in the fluvial environment, as already mentioned, there are no references in the literature about calculations of emissions, or environmental or social impacts. Therefore, the main objective of this article is to modify the methodology developed to study the maritime environment [9] so that it is possible to study the fluvial environment.

2. MATERIAL AND METHODS

All the procedures indicated below are a modification of the methodology for the analysis of maritime cruise traffic developed by Vicente-Cera et al. (2019) [9], applied to the case of river cruises. Please refer to the original citation for further details of the methodology as well as the steps not indicated in this article, as only those steps with significant modifications will be described.

The main difference between the two methods is that the calculation of navigation distances in rivers presents particular challenges with respect to the calculation of navigation distances in oceans that must be solved, the main objective of this article.

2.1 ASSOCIATION OF AIS DATA WITH RIVER BASINS

In the original methodology AIS data were associated with marine regions of the global ocean, and in the present study a modification was made, associating AIS data with river basins. To be able to make comparisons of cruise traffic in the different river basins containing different areas of influence, size and different navigable watercourses, it is necessary to be able to refer AIS data to the different river basins present in the world. However, the ten main rivers with relevant cruise traffic (mainly in Europe and the United States) accumulate 96% of the hours navigated in rivers, so only cross data were kept for these basins.

Operationally, to make the association of AIS records to each body of water was used the software GIS QGIS®, specifically the tool “join attributes by location”, which from the latitude and longitude of the record AIS assigns each point to the area of the river basins that corresponds with its associated variables. The base layer containing the polygons with the hydrographic basins of the world is that provided by the Geonetwork programme of the Food and Agriculture Organization of the United Nations (FAO) [12]. AIS data come from a database acquired through VesselFinder® containing 163,222 records for the year 2016 and with a time resolution of 12h and worldwide spatial resolution.

2.2 PROCEDURE FOR CALCULATING NAVIGATION DISTANCES IN RIVERS

When measuring navigation distances with the Haversine approach as mentioned in [9] the main problem is that distances are measured by the approximation of large arcs, i.e. the arc of the earth's surface that joins the two points. However, when this approximation is used in rivers the errors made are very large since rivers tend to have sinuous rather than straight shapes, so that the distance between point A and B is not the arc that joins the two points but mandatorily the entire length of the river that links these points.

To solve this problem it is necessary to have a tool capable of connecting the route of the AIS registers along the course of the river. Operationally the QGIS® software was used to carry out this complex procedure, starting with the tool "points along the geometry", which will allow us to divide the linear layer of the river into segments of equal size distributed along its entire length and which generates a new layer called "Distance". It is used as division rate 0.01 units, which is equivalent to about 850 meters (since the base layers used are georeferenced in degrees). The base layer with information from the most relevant rivers in the world was obtained from Natural Earth (2018) [13].

Subsequently, through the "join point to point" tool applied between the AIS data layer and the "Distance" layer, it generates a new layer called "Result" that will be associated with each of the AIS records to one of the previously obtained points for the river in question (the closest one) as shown in Figure 1. In this way, the AIS records data will be associated to a specific distance from the river head. Finally, the "join attributes by location" tool applied between the "Result" layer and the AIS layer will generate a layer called "Join Layer" in which the river distance data will be included in the AIS attributes table.

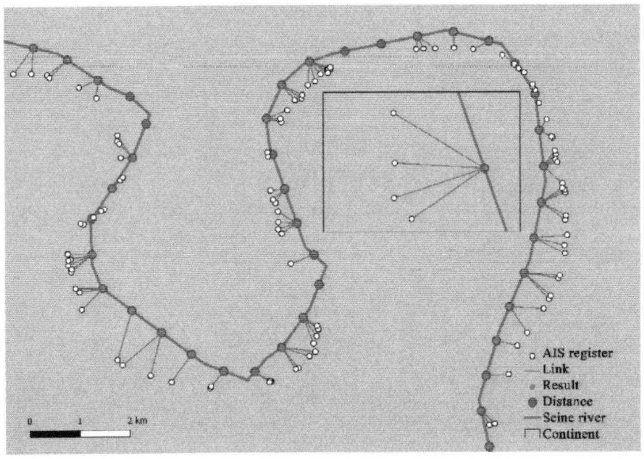

Figure 1 Distance calculation process in the Seine River (France). The red dots correspond to the equidistant distance segments of the river. White dots with AIS records. The green dots are the result of assigning an AIS point to a distance point, and the red line the movement made. Several AIS records associated with the same distance point are shown in detail.

To estimate the distances travelled between two AIS points, the distance attributes of both points will be subtracted and multiplied by the value of each distance unit. This distance unit is found by dividing the length of the river (taken from the river information layer) by the total number of units into which the river was divided. In this way it is easy to calculate how much distance a cruise has travelled between two points of the river, in a much more precise way, by crossing these points unequivocally.

3. RESULTS AND DISCUSSION

Analysing AIS data for the world fleet of river cruise ships: in 2016 it had 296 active ships, 2,546,801 passengers, and a total navigated distance of 5,556,177 km. The oldest ship still in operation dates from 1938, while the most modern in the time series had an expected launch in early 2017 (available data until 2016). Figure 2 shows the growth in tonnage of these ships. Although exponential in the early years, it soon stopped in a range of sizes that seems to have reached its limit, placing the average tonnage of the fleet at 3,059 GRT. With 113m in length and 14m in beam, considering that the largest cruise ships currently sailing the rivers of the world have lengths not exceeding 140m.

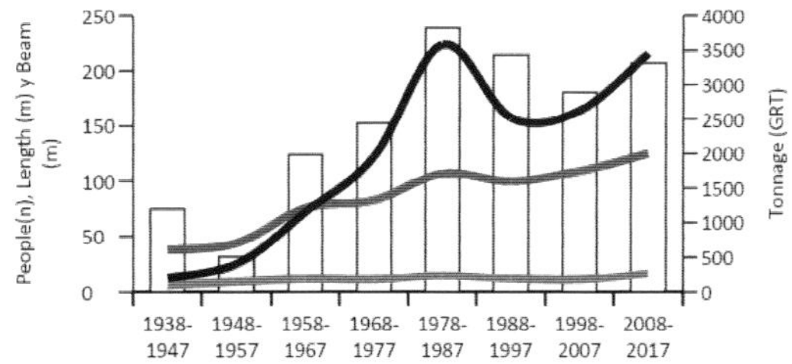

Figure 2 Evolution of the fleet of river cruise ships worldwide. The main axis represents the average number of people (white bars), length (red line), and beam (grey line) of the ships. The secondary axis represents the average tonnage (black line).

This limitation in the growth of ships is imposed by the characteristics of the river's contours, preventing an increase in the beam, which has remained relatively constant over the last eighty years. On the contrary, the length of vessels has been increasing as the demand for these vessels has been greater, however, the growth of this also seems to have slowed over time. The only growth option for this type of tourism is the launching of new ships. As can clearly be seen in Figure 3, the number of ships of this type of tourism has grown exponentially as opposed to their size. A characteristic that defines river cruises and distinguishes them completely from the oceanic ones.

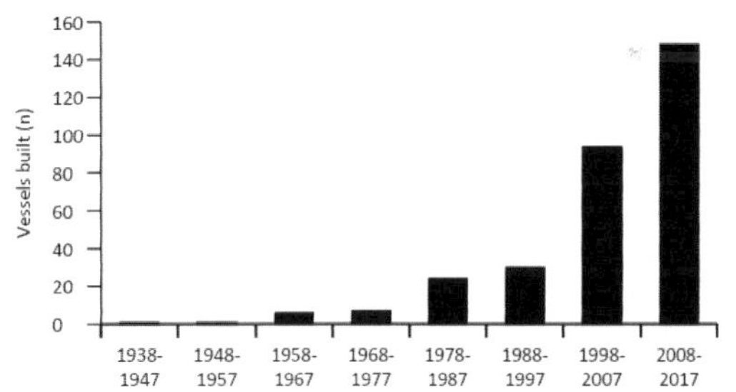

Figure 3 Number of river cruise ships built per year over the period studied.

On the one hand, this reduces emissions in the sense of pollution per ship unit, however, it subjects to greater environmental pressure those rivers with greater tourist, historical, and natural interest. Thus, in some cities it is possible to see "caravans" of these ships sailing in formation with the aim of accommodating as many passengers as possible, including new forms of impact on the river environment: the visual and landscape impacts. There are even examples in the literature of how cruise traffic itself in certain areas is capable of modifying the distribution of riverine vegetation and the surrounding ecosystem [14].

3.1 GLOBAL PRODUCTION OF WASTE AND EMISSIONS

The ten basins most navigated by river cruise traffic constitute 94% of the navigation hours of the entire fleet. If we add to this other basins of interest such as those of the Mississippi, Columbia, and Guadalquivir rivers, it results in a navigation time percentage of 95%. That is why studying these basins in detail is an approximation of the global impact that this industry generates. Figure 4 shows how Central Europe is the region with the highest concentration of this type of tourism, highlighting the Rhine basins (38,283 navigated days year^{-1}; 52%), Danube (25,671 navigated days year^{-1}; 35%) and Seine (2,512 navigated days year^{-1}; 3%).

In the vast majority of navigable rivers, the dumping of waste water, bilge, waste, and other pollutants is completely prohibited, so these have to be managed in port. As a result of this the cruise ship must make regular stops in cities equipped with these facilities in addition to having sufficient storage for the waste they produce [15,16]. This can be observed when checking the times of the ship, dedicating almost half of its activity to being stopped at port (32%), and the rest being dedicated to navigation (68%).

Table 1 Global waste generation, emissions and shipping statistics from the river cruise industry (2016). The most important river basins of the planet are represented.

River basins	Navigated time (d)	Navigated distance (km)	Passenger (n)	Paths (n)	Black waters (m^3)	Gray waters (m^3)	Waste waters (m^3)	Bilge water (m^3)	Fuel (kg)	CO_2 (kg)	Waste (kg)
Rhine	38,283	1,763,978	1,347,727	8,707	394,587	970,601	1,365,188	21,681	23,150,774	94,016,742	57,773,060
Danube	25,670	3,121,079	826,021	5,048	272,576	670,478	943,054	38,361	42,198,909	171,372,412	39,908,870
Rhône	1,387	220,907	157,289	1,027	13,589	33,426	47,015	2,715	2,755,823	11,191,569	1,989,600
Seine	2,512	247,497	146,401	1,062	23,458	57,702	81,160	3,042	2,942,919	11,951,380	3,434,592
Scheldt	856	41,549	120,608	617	9,368	23,043	32,411	511	578,985	2,351,294	1,371,600
Volga	2,144	43,999	114,026	542	28,700	70,596	99,296	541	749,956	3,045,616	4,202,083
Duero	988	18,060	58,354	433	8,561	21,058	29,620	222	199,261	809,211	1,253,464
Columbia	353	29,164	37,219	198	4,267	10,496	14,763	358	448,883	1,822,940	624,756
Mississippi	225	34,614	29,394	93	4,392	10,803	15,195	425	860,308	3,493,763	643,032
Meuse	148	9,364	24,628	203	1,422	3,497	4,919	115	114,534	465,131	208,153
Elbe	986	42,791	21,733	307	6,532	16,067	22,598	526	360,944	1,465,816	956,338
Guadalquivir	130	1,207	12,144	69	1,485	3,652	5,137	15	17,552	71,278	217,384
Total	**73,682**	**5,574,209**	**2,895,544**	**18,306**	**768,937**	**1,891,419**	**2,660,356**	**68,512**	**74,378,847**	**302,057,153**	**112,582,931**

However, beyond the pollutants that can be discharged into the river, certain amounts of CO_2 are inevitably emitted into the atmosphere (302 kt year^{-1}), usually in the environment of densely populated cities with heavy road traffic, which can contribute to worsening air quality in these environments. The regulations on the fuel these vessels must use are strict, so sulphur emissions are limited by always using low sulphur fuel (ULSD). By the end of 2050 all vessels in the world must use this type of fuel [17]. This is why NO_x, SO_x, and $PM_{2.5}$ parameters have not been included in the results of river cruise ships, since they seem not to be produced with this type of fuel or they are of very low draft.

In terms of wastewater, this form of tourism only generates around 2.7 hm^3 yr^{-1} worldwide, which is a small amount if we distribute it among all the cities and locations through which they circulate. However, we must consider that the concentrations of pollutants found in cruise waters are really worrying [18]. Managers of purification facilities must make an effort to know which waters they are receiving and whether they are being properly treated before being returned to the river, since theliterature detailing the presence of the pollutants mentioned in this work along numerous rivers in the world is extensive [19, 20]. Inappropriate management of these waters can end up having an impact on the river waters, although the discharge does not occur directly from the ship, producing localized pollution in the course of the river and transporting pollutants from one point to another, even reaching the sea or persisting in river dynamics [20].

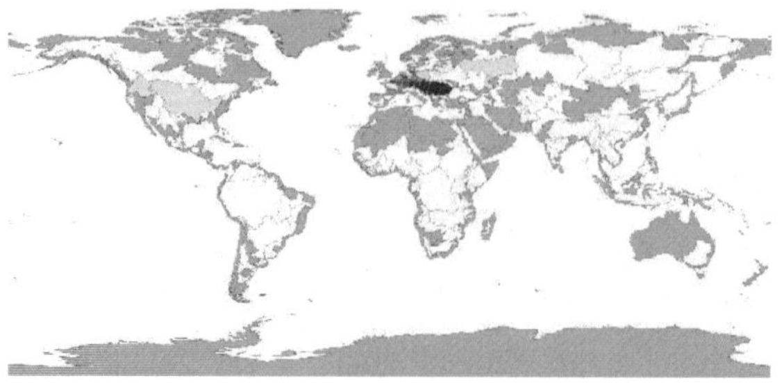

Figure 4 The most transited river basins by river cruise ships at global level, expressed in number of AIS records per basin.

3.2 COMPARISON OF OCEAN AND RIVER TRAFFIC

Figure 5 shows a heat map of world cruise traffic, both fluvial and oceanic, showing how this type of tourism affects the entire planet. This map shows that there are practically no populated coasts on which this traffic has not already appeared, with an important boom in the Asian market that will be commented on later. In the case of navigable rivers, although new possible destinations are beginning to appear, as is the case of the Yangtze River, these are still very restricted to classic destinations, especially in Central Europe.

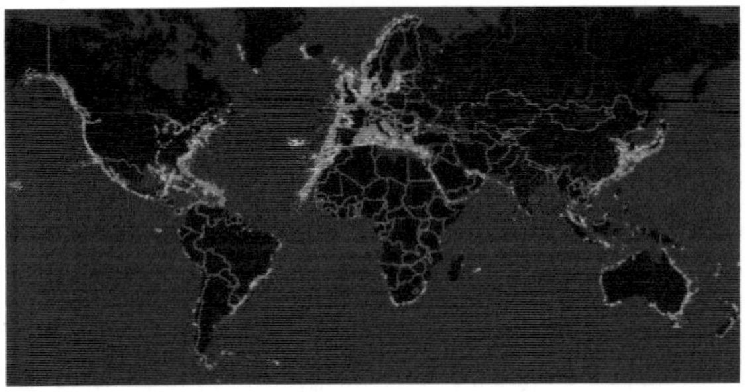

Figure 5 World heat map of cruise ship traffic (river and ocean). This map from blue to red shows the areas with the most cruise traffic.

As has been described throughout this paper, this type of tourism of undeniable social, recreational, and economic value has the counterpart of an enormous associated pollution, which must be known and tackled, not by prohibiting or limiting its activity, but by finding strategies that make this means of transport, tourism, and leisure sustainable in terms of its impact on the environment. Table 2 shows the total waste produced by the entire cruise industry, both fluvial and oceanic, revealing the enormous amounts of CO_2 that are handled, in addition to wastewater and associated waste.

Although both industries have a similar number of ships, 369 in the case of the oceanic and 296 for the fluvial, these vessels are very different in terms of size. Although in both cases the lengths are similar on average (197m oceanic; 113m fluvial) they are radically different in tonnages: oceanic vessels handle average sizes of 54,000 GRT and a maximum size of 227,000 GRT; while fluvial vessels handle average sizes of 3,000 GRT with a maximum of 7,000 GRT. It should also be noted that both types of vessels operate differently, e.g. oceanic vessels move on average at 10 kn (knots), while river vessels move only at 2 kn.

Table 2 Basic statistics on the size of cruise ships, both fluvial and oceanic.

		Ocean	*River*
Passengers (n)	*Min*	40	20
	Mean	1.493	157
	Max	6.224	260
Length (m)	*Min*	10	17
	Mean	197	113
	Max	371	136
Beam (m)	*Min*	7	4
	Mean	27	14
	Max	72	29
Tonnage (GRT)	*Min*	140	30
	Mean	54.751	3.060
	Max	227.700	6.916
Age (year)	*Min*	1.931	1.938
	Mean	1.997	2.003
	Max	2.017	2.016
Tonnage per passenger (GRT pax^{-1})	*Min*	1	1
	Mean	40	15
	Max	1.690	63

*Min: Minimum; Max: Maximum.

The proportion of waste, discharges, and emissions for both fleets illustrates this as it is evident that ocean traffic is more worrying in numbers as it accumulates an average of 94% of the production of waste discharges and emissions while the river generates on average only the remaining 6% (Table 3). However, this must be treated with caution as the dynamics of the two environments are really different and it is necessary to study in detail how the pollutants behave in each of these environments.

Table 3 Global waste production, emissions and shipping statistics from the cruise industry, both river and ocean (2016)

	Ocean	*%*	*River*	*%*
Active vessels (n $año^{-1}$)	369	*55*	296	*45*
Port calls (n $año^{-1}$)	64,323	*37*	107,986	*63*
Navigated distance (km $año^{-1}$)	36,657,852	*87*	5,574,209	*13*
Passengers (n $año^{-1}$)	25,508,221	*90*	2,895,544	*10*
Navigated time (d $año^{-1}$)	82,771	53	73,682	*47*
Port time (d $año^{-1}$)	71,067	*67*	34,304	*33*
Grey waters (hm^3 $año^{-1}$)	22,5	92	1,9	*8*
Black waters (hm^3 $año^{-1}$)	9,31	92	0,8	*8*
Wastewater (hm^3 $año^{-1}$)	31,8	92	2,7	*8*
Bilge waters (hm^3 $año^{-1}$)	0,46	*87*	0,1	*13*
Fuel (kt $año^{-1}$)	6,018	*99*	74	*1*
CO_2 (kt $año^{-1}$)	21,391	*99*	302	*1*

Waste (kt año^{-1})	1,977	*95*	113	*5*

3.3 ECONOMIC IMPACT OF THE SECTOR

From the economic data provided by the Cruise Market Watch® [2] platform and the data obtained in the present work on the number of passengers mobilized it is possible to form an economic balance of the activity of the cruise industry both oceanic and fluvial.

In economic terms the oceanic traffic of cruise ships in 2017 generated an economic movement for the industry of €32,000 million year^{-1}, working on the basis that the average spend of a passenger on board ship is €224 passenger^{-1} day^{-1} [21]. And knowing that the average expenditure in a destination of a passenger of this type of tourism is around €75 passenger^{-1} destination^{-1} [22,23], the economic movement that occurs worldwide to the different destinations is €9,000 million.

The ten destinations with the highest revenues in 2017 were: Cozumel (Mexico) (€281 million year^{-1}), Nassau (Bahamas) (€255 million year^{-1}), Miami (United States of America) (€186 million year^{-1}), Shanghai (People's Republic of China) (€160 million year^{-1}), Barcelona (Spain) (€155 million year^{-1}), Cape Canaveral (United States of America) (€140 million year^{-1}), Civitavecchia (Italy) (€131 million year^{-1}), Fort Lauderdale (United States of America) (€110 million year^{-1}), Marseille (France) (€102 million year^{-1}), and Palma de Mallorca (Spain) (€100 million year^{-1}). All of them add up to a total of €1,620 million, which results in 18% of the total world profit, so the rest of the destinations accumulate a total of €7,380 million (82%) but this is much more widely distributed throughout the world.

In total the global economic movement generated by the oceanic sector of the industry, if we add up everything generated for the industry (€32,000 million year^{-1}) and that generated in all destinations in the world (€9,000 million year^{-1}), results in €41,000 million year^{-1}. To these figures should be added other secondary costs that this industry can generate: the cost of repairs, construction of the ship, supplies of water, drink and food, chemicals and cleaning products, jobs, and so on.

On the other hand, the fluvial traffic of cruise ships (Table 4), considering an average expense per passenger of €517 person^{-1} day^{-1} [24], is considerably higher than the cost of an average oceanic cruise. This industry generates profits of 8,955 million euros per year. Add this to the economic expenditure generated in the destinations they visit, again using an average expenditure of €75 passenger^{-1} destination^{-1} [22,23] results in a profit for the different destinations of €1,300 million year^{-1}.

The basins that derive most benefit are: the Rhine (€617 million year^{-1}), the Danube (€465 million year^{-1}), the Seine (€60 million year^{-1}) and the Volga (€37 million year^{-1}); with special mention being made of the basins of the Iberian Peninsula: Duero (€17 million yar^{-1}), and Guadalquivir (€5 million year^{-1}).

It is important to note that one sector, that of river cruise ships, which transports barely 10% of cruise users with respect to the oceanic sector, accumulates around 20% (€8.955 million year^{-1}) of the world economic movement. This is probably due to the fact that the prices of river cruises are significantly higher (€500 day^{-1} versus €200 day^{-1}) and also that the costs of operation and construction of new ships in both sectors are radically different.

While oceanic cruises have become popular generating a wide and varied offer, which leads to an inevitable drop in prices, river cruises can still be considered as a form of luxury tourism, for a more specific public and greater purchasing power that pursue a different experience to mass tourism.

Table 4 Economic movement generated by the river cruise industry in the world's most navigated basins.

River basins	*Shipping profit (€ año^{-1})*	*Destination profit (€ año^{-1})*	*Mean Fare (€ pax^{-1})*
Rhine	4,249,809,306	616,510,054	3,073
Danube	3,208,478,516	465,446,593	3,884
Seine	418,051,589	60,645,782	2,735
Volga	256,749,495	37,246,058	1,976
Rhône	234,051,715	33,953,344	1,615
Duero	113,839,797	16,514,477	1,969
Scheldt	111,426,623	16,164,404	1,071
Elbe	72,536,764	10,522,741	3,336
Columbia	168,002,037	24,371,669	4,555
Mississippi	57,993,613	8,413,000	2,073
Meuse	32,927,629	4,776,735	1,096
Guadalquivir	31,752,417	4,606,250	2,615
*Total/ Mean**	8,955,619,501	1,299,171,108	2,500*

Together, both sectors (ocean and river) accumulate a world economic movement of €51,000 million year^{-1}, which affect a large percentage of shipping, but also the destinations and regions that are visited. This is most definitely an industry that is an economic and employment engine which deserves to be studied in detail so that its activities continue in a sustainable way, compatible with humanity's commitment to preserve the environment.

7. CONCLUSIONS

In the last eighty years river tourism has grown in parallel to the oceanic tourism in number of vessels. However, the dimensions of these vessels are limited by the draft of the waters they navigate, so although their size has increased significantly it has stagnated in recent years, mainly due to the limiting factor of river space.

Since the 1970s the number of ships making up the fleet of river cruise ships has increased exponentially, from a few dozen in the early years to a peak in the last decade with more than 296 active ships.

It has been observed that the Haversine formula is not appropriate for measuring distances in the river environment, mainly due to the sinuous shape of the river. The process developed through GIS techniques has proved to be more reliable when calculating distances by ships in rivers, obtaining better results in those rivers with more complex shapes.

The ten most active cruise ship river basins in the world account for 96% of the navigation hours of this industry sector. Moreover, the first three basins (the Rivers Rhine, Danube, and Rhone) alone account for 77% of the activity. So it is a tourism that is really concentrated in a few river basins.

Although both ship industries (river and ocean) affect their respective environments with their own environmental pressure, globally the maritime industry accounts for between 87 and 99% of discharges, waste, and emissions from cruise ships (94% on average), while the river varies in its parameters of pollutants between 1 and 13% (6% on average).

The global cruise ship industry produces a total economic turnover of €50,000 million $year^{-1}$ (€41,000 million $year^{-1}$ for the oceanic industry, and €9,000 million $year^{-1}$ for the river industry), which places it among one of the most productive industries not only in the tourism sector, but also in the entire commercial world.

8. ACKNOWLEDGEMENTS

This work was supported by the Spanish Ministry of Economy and Competitiveness, AVANTE project [CTM2014- 52116-R, 2014]

9. REFERENCES

[1] European Environmental Agency (EEA). Aviation and shipping — impacts on Europe's environment TERM 2017: Transport and Environment Reporting Mechanism (TERM) report. 2017. https://doi.org/10.2800/4907.

[2] Cruise Market Watch. 2015 Cruise Trends Forecast. Cruise Mark Watch 2018. http://www.cruisemarketwatch.com/ (accessed July 17, 2018).

[3] Ward D. Berlitz Cruising & Cruise Ships 2016. 2nd ed. Berlitz Travel; 2015.

[4] Ward D. River cruising in Europe. London: Berlitz; 2016.

[5] Quesada R. Elementos de Turismo. Teoría, Clasificación y Actividad. 2nd ed. EUNED; 2010.

[6] Davies EAJ, Marsh CM. Canals and inland waterways. Encycl Br 2013. https://www.britannica.com/technology/canal-waterway (accessed July 17, 2018).

[7] Santiso A. Historia de los canales. Tur Fluv 2018.http://www.turismofluvial.com/historia-canales.php (accessed July 17, 2018).

[8] Whitmore M. Transport and supply during the first world war. Imp War Museums 2018. https://www.iwm.org.uk/history/transport-and-supply-during-the-first-world-war (accessed July 17, 2018).

[9] Vicente-Cera I, Acevedo-Merino A, López J, Nebot E. Use of AIS data for the environmental characterization of world cruise ship traffic. Int J Sustain Transp 2019:1–10. https://doi.org/10.1080/15568318.2019.1575494.

[10] Jalkanen JP, Johansson L, Kukkonen J. A comprehensive inventory of ship traffic exhaust emissions in the European sea areas in 2011. Atmos Chem Phys 2016;16:71–84. https://doi.org/10.5194/acp-16-71-2016.

[11] Wu L, Xu Y, Wang Q, Wang F, Xu Z. Mapping Global Shipping Density from AIS Data. J Navig 2016:1–15. https://doi.org/10.1017/S0373463316000345.

[12] Aguilar-Manjarrez J. WRI Major Watersheds of the World Delineation. FAO-GeoNetwork 2006. http://www.fao.org/geonetwork/srv/en/main.home?uuid=8a3f1390-e28c-11db-a939-000d939bc5d8 (accessed November 26, 2018).

[13] Natural Earth. Rivers and lake centerlines 2018. https://www.naturalearthdata.com/downloads/10m-physical-vectors/10m-rivers-lake-centerlines/ (accessed November 8, 2018).

[14] Ali MM, Murphy KJ, Langendorff J. Interrelations of river ship traffic with aquatic plants in the River Nile, Upper Egypt BT - Biology, Ecology and Management of Aquatic Plants. In: Caffrey J, Barrett PRF, Ferreira MT, Moreira IS, Murphy KJ, Wade PM, editors., Dordrecht: Springer Netherlands; 1999, p. 93–100.

[15] Grammerstorf HH. European River Cruising and ecological aspects of river cruises 2013.

[16] The Travel Foundation. Environmental Sustainability for River Cruising. 2013.

[17] Beecken J, Mellqvist J, Salo K, Ekholm J, Jalkanen JP, Johansson L, et al. Emission factors of SO<inf>2</inf>, NO<inf>x</inf> and particles from ships in Neva Bay from ground-based and helicopter-borne measurements and AIS-based modeling. Atmos Chem Phys 2015;15:5229–41. https://doi.org/10.5194/acp-15-5229-2015.

[18] Vicente-Cera I, Moreno-Andrés J, Amaya-Vías D, Biel-maeso M, Pintado-herrera MG, Lara-martín PA, et al. Chemical and microbiological characterization of cruise vessel wastewater discharges under repair conditions. VI Int. Symp. Mar. Sci., Vigo (Spain): 2018. https://doi.org/10.13140/RG.2.2.25308.28800.

[19] Wilson B, Chen RF, Cantwell M, Gontz A, Zhu J, Olsen CR. The partitioning of Triclosan between

aqueous and particulate bound phases in the Hudson River Estuary. Mar Pollut Bull 2009;59:207–12. https://doi.org/10.1016/j.marpolbul.2009.03.026.

[20] Sun Q, Li Y, Li M, Ashfaq M, Lv M, Wang H, et al. PPCPs in Jiulong River estuary (China): Spatiotemporal distributions, fate, and their use as chemical markers of wastewater. Chemosphere 2016;150:596–604. https://doi.org/10.1016/j.chemosphere.2016.02.036.

[21] Cruise Market Watch. 2018 Cruise Trends Forecast. Http://CruisemarketwatchCom 2018.

[22] Business Research & Economic Advisors. Aportación económica del turismo de cruceros en México 2007;I:225. https://doi.org/10.1017/CBO9781107415324.004.

[23] Brida JG, Velásquez CMG, Aguirre SZ. Un análisis del gasto del pasajero de cruceros que visita Costa Rica. Anu Tur y Soc Univ Extern Colomb 2010;11:35–50.

[24] Avid Travel Media. River Cruise Price Comparisons. River Cruise Advis 2018. https://www.rivercruiseadvisor.com/river-cruise-price-comparisons/ (accessed November 27, 2018).

SILENT VESSELS TECHNOLOGY AND THE REDUCTION OF UNDERWATER RADIATED NOISE AS A SIGNIFICANT ISSUE FOR THE FUTURE OF MARITIME ENGINEERING

P. Beltrán, Técnicas y Servicios de Ingeniería (TSI SL, publiobp@tsisl.es), Spain

SUMMARY

Underwater radiated noise (URN) due to shipping activity has recently received considerable attention in the scientific community because of its negative environmental impact on marine fauna. High levels of URN emitted by vessels block the animal's ability to communicate, reproduce, navigate or hunt, which is vital for the sustainability of the oceans. As a result, new regulatory frameworks to control the acoustic signature of vessels are being developed and promoted by IMO to mitigate the negative effects of URN in the oceans, while, simultaneously, these actions can be used to reduce the noise on the vessel and improve the comfort of the crew on board.

This study describes a validated methodology applied to the most modern Fishing Research Vessels (FRVs), the most restricted civilian vessels in terms of noise radiated to the water, to minimize their underwater acoustic signature, which is mainly originated by propellers and main machinery. To achieve this goal, an extensive analysis using advanced simulation software and experience is performed to predict the noise and vibration (N&V) emissions of the main noise sources in the ship. Based on the results of this analysis, preventive control actions are applied to the noise sources with the aim of minimizing the emission of URN in the oceans.

Findings obtained over the years reveal that this procedure can be successfully used to predict and reduce the URN generated by any type of ship. Based on the author's experience, this cutting-edge solution based on N&V comprehensive management analysis should be performed from the ship design phase to detect acoustic problems in the early stages of the project and to control them till the end of the construction process to minimize costs and reduce URN negative impacts in the oceans.

NOMENCLATURE

URN	Underwater Radiated Noise
FRV	Fishing Research Vessel
IMO	International Maritime Organization
MEPC	Marine Environment Protection Committee
MSFD	Marine Strategy Framework Directive
N&V	Noise and Vibration
FEA	Finite Element Analysis
BEM	Boundary Element Method
SEA	Statistical Energy Analysis
CFD	Computational Fluid Dynamics
NRH	Noise Radiated to the Harbour
N&VCM	Noise & Vibration Comprehensive Management
ICES	International Council of Exploitation of the Sea
DCPA	Distance of Closest Point of Approach

1. INTRODUCTION

As is well known by everyone, the UN-Climate Change Conference COP25 took place in Madrid from 2nd to 13th December 2019, with a large media deployment for the politicians and their "young and conjunctural mediatic instruments". Despite of the recognition of the coming environmental risks of the planet with the current growing model, there was no consensus related to the actions to be taken. As can be seen through the published documents, all conferences have focused their attention, which is important and critical, on the "visible pollutant agents", but nothing has been said about the "silent enemy" of the marine life in the oceans. In relation to this, it is important to mention that during a round-table addressing younger generations, only a Spanish SME, which represents the author, has tabled this issue.

In short, this was a missed opportunity to raise awareness in society of a problem that the Scientific and Biological Communities have been demanding for decades, and which, as recognized by the IMO-International Maritime Organisation, itself a UN public agency, is a "global problem that requires international collaboration and urgent solutions". In short, there is nothing strange in politicians' behavior: they only deal with problems when they explode in their hands, or in the future that future generations will occupy.

Despite this situation and the observations made to us by some managers of reputable maritime organizations, as engineers, not ecologists, fully convinced that the current great challenge of all the branches of science and engineering is to find solutions that make the development of societies compatible with environmental sustainability, we will continue our fight, of more than forty years, against the underwater "silent and invisible enemy" radiated by ships.

Within the still "confusing scenario" of URN that will be detailed and commented on later, this paper mainly aims to present the tool that, based on the experience gathered over forty years of his professional life, has been used by the author to design and to build "silent ships" in compliance with the most strictest requirements of not only noise and vibration on board, noise radiated to the harbour, but also underwater radiated noise (URN). This tool, called "Noise & Vibration Comprehensive Management" (N&VCM) is a clear and practical example of how some owners and shipyards, within the shipbuilding industry, are aware of this new topic of the URN and how they are moving ahead to reduce the environmental impact of new ships.

This tool, improved a long time, has been and is being applied, in different grades, in the dynamic and acoustical design of Fishing Research Vessels as well other vessels such as luxury yachts, passenger ships, ferries, naval ships, and all kind of civilian ships. Its validation, together with the experimental results obtained during the sea trial measurements, confirmed to us that we are in front of a proven and efficient tool that will enable us to control and abate the impact of the ships not only for their passengers and crews, the surrounding areas of the harbours, but also for the marine life by the reduction of underwater noise radiated by the ships.

In this last aspect, related to the reduction of the URN by ships, special mention should be made to the successful results obtained by the application of the N&VCM to the dynamic and acoustical design of seven (7) FRVs. The FRVs are the only civilian vessels that must meet, by performance requirements, the strict ICES CRR 209 requirement of underwater radiated noise.

Finally, two main conclusions could be drawn from this paper. Firstly, marine engineering has proven and validated tools that enable the successful tackling of the most recent challenge of the shipbuilding industry, that is the reduction of the noise radiated in the water by all type of commercial ships. The solution is just a matter of will and decision-making. Secondly, the most modern FRVs that have achieved the compliance with the requirements of ICES CRR209 become the best technological references for the procedure to be followed to achieve the target: reduction of the underwater radiated noise by all type of commercial ships.

1. THE CURRENT CONTEXT

1.1 TREND OVERVIEW AND REGULATORY SCENARIO

In the last two decades the different policies of the European Commission (EC) within the framework of the Green Policy have been focused on combating and reducing the current environmental impact of all types of vessels that make up the current European Fleet. Other international institutions such as IMO and ILO are moving in the same direction. To fulfil this main target and engage companies and regulatory bodies, specific R&D projects [1] have been promoted and launched by the EC. A common conclusion arising from all these projects is that the N&V-Full Signature (N&V-FS) of the vessel is required to assess properly its environmental impact. This key indicator consists of (i) the noise and vibrations (N&V) on board, (ii) the airborne noise radiated to the harbour (NRH), and (iii) the underwater radiated noise (URN), being URN the greatest challenge to be faced by the civil shipbuilding industry, because if a vessel has a low URN emission, N&V and NRH will be automatically low.

Within this regulatory trend, the European Union is considering different Directives such as 2003/10/EC (to protect the health of all seamen without exceptions against the noise exposure levels), and Directive 2006/56/EC (to avoid the annoying acoustical impact of the ships in the residential areas surrounding harbours) among others, which have been released and extrapolated to EU Member States as mandatory local laws.

The results achieved by all players involved, including the current author, during decades of efforts in the abatement of the noise in the marine sector, are finally tangible. Nowadays the URN from commercial shipping has definitively been admitted by public institutions and regulatory bodies, as an anthropogenic pollution that may have adverse impact on the marine environment and so have brought short and long-term negative consequences on marine life, especially marine mammals. Although different international environmental conventions and directives have tried, for many years, to address the different noise polluters, the continuous noise from commercial shipping remained out of their scope until 2008.

The Marine Strategy Framework Directive (MSFD) [2], issued in 2008, recognized that all efforts made to preserve the biodiversity of the EU marine environment had to include the anthropogenic underwater noise pollution. This directive aims at defining and reaching a Good Environmental Status (GES) in all EU waters by 2020.

Shipping is admitted to be one of the largest contributors to the low frequency ambient noise in the oceans. The issue of URN was first raised at International Maritime Organization (IMO) in 2004. The Marine Environment Protection Committee (MEPC) on Noise Impact on Marine Life has finalized the Non-Mandatory "Guidelines for the Reduction of underwater noise from commercial shipping to address adverse impacts on Marine Life" [3] with its approval in April 2014. Despite the non-mandatory character of these guidelines, they include in the chapter 5 (Predicting underwater noise) different engineering and simulation tools that may assist in this task. Among others and according to the author's experience, it is worth mentioning the following ones: (i) using CFDs for improvement of the wake field of the propeller as well as predicting propeller cavitation, and other phenomena produced by hydro-dynamic noise such as bubble sweep-down circulation along the hull; and (ii) Finite Element Analysis (FEA), Boundary Element Method (BEM) and Statistical Energy Analysis (SEA), for the simulation of the noise transmitted from the machinery to the water through the structure of the ship.

1.2 METHODOLOGICAL AND TECHNICAL STANDARDS FOR MEASUREMENT AND ASSESSMENT OF THE SHIPS URN

It is important to mention that today there is still a lack of agreement between main stakeholders about methodological and technical standards for the measurement of the URN of ships and their corresponding assessment. It cannot be forgotten that the characterization of each ship as a source of noise radiated to water and its harmonization is essential from the point of view of noise control engineering [4]. Currently, the methodologies and measurements procedures for civilian ships could be classified into two different groups: (i) those coming from international standardisation bodies, mainly addressed to the noise measurement procedures without any noise limits for assessment; and (ii) those released by the Classification Societies as URN Class Notations with their own noise limits and measurement procedure.

An essential parameter is transmission loss (TL), used to calculate the radiated noise at 1m at both sides of the ship from the noise signal recorded at the closest distance between the target ship, according to the defined test speeds and the hydrophones hanging line. Real transmission loss depends on several parameters and is usually estimated. It represents a current uncertainty that must be harmonized, because if it the same TL is not used, obviously the URN values obtained for the same vessel and recorded noise signal would be different.

Below is summarized several measurements procedures and methodologies regarding URN measuring.

1.2 (a) ANSI/ASA-2009

As a first try to standardize the URN measurements procedure, ANSI-ASA released its standard "Quantities and Procedures for Description and Measurements of Underwater Sound from Ships, Part 1: General Requirements" [5]. This addresses measurements of the underwater sound pressure in the far field and deep waters to avoid reflections. The procedure selected has no inherent limitation with regards to the vessel size and mainly considers the hydrophone array connected to the seabed. The result of the measurement is the noise pressure levels in third-octave in decibels (dB) referenced to 1µPa at 1 m from both hull sides. Band levels are associated with the mid-frequency of the third-octave band. ANSI-ASA does not define targeted limits

1.2 (b) ISO-17208-1

In 2016 ISO (International Standard Organization) released the standard "Underwater Acoustics-Quantities and procedures for description and measurement of underwater sound from ships, Part 1: Requirements for precision measurements in deep water used for comparison purposes" [6]. This specifies the general measurement system, procedure, and methodology used for the measurement of URN from ships under a prescribed operating condition.

The resulting figures are based on the root-mean-square sound pressure levels (SPL), herein used synonymously with sound pressure level or SPL measured in the far field of the ship and normalized to a distance of 1m and reported in one-third octave bands. The result of these measurements is called "radiated noise level". The method is limited to ships moving at speeds no greater than 50 knots (25,7 m/s). The measurement method smooths the variability caused by Lloyd's mirror surface image coherence effects but, nevertheless, does not exclude a possible influence of propagation effects like bottom reflections, refraction, and absorption. No specific adjustments for these effects are provided in this part of this standard which is solely focused on measurement in deep waters. The intended uses of the method are: (i) to show compliance with contract requirements, (ii) for vessels comparison, (iii) to periodic acoustic signature assessments, and (iv) for research and development activities. Deviations around 10 dB (ref. 1mPa @ 1m) have been obtained from several tests for the same ship in shallow water and deep water.

Following the discovery of the limitations of this method, ISO announced the second part (ISO-17208-2) of this standard (measurements in shallow water, under revision), which is expected to be released in 2019.

It is worth remarking that the requirement of deep water to minimize the effect of the seabed will bring great difficulties for their application in areas with shallow water such as the Adriatic Sea, the Baltic Sea, the North Sea, and many other areas of Asia and the rest of the world.

1.2 (c) ICES CRR Nº209

This represents the URN regulation of Fishing Research Vessels (FRV) in the framework of civil ships and it was released in 1995 by ICES (International Council of Exploitation of the Sea) [7]. Based on more than forty years of biological research of this institution, it recommended the URN target limits detailed in the Figure 1 from 10 Hz to 100kHz. These limits are expressed in term of 1/3 octave band spectrum levels referred at 1 Hz, at 1m of both hull sides and at an operational speed of 11 knots. These limits, with two different slopes for two specific frequency ranges, were defined to ensure optimal performance by avoiding the disturbance of the target fish species subject to observation (limit in the frequency range from 10 Hz to 1kHz) as well as the interferences in the echo-sounders (from 1kHz to 100 kHz).

The positive effects of its implementation in the segment of FRVs have been: (i) inclusion of a specific clause of strict compliance with the URN limits in the contracts with shipyards by shipowners of all the modern research vessels; (ii) implementation of significant changes and improvements in main propulsion systems; and, (iii) the lack of available experimental data, emphasized and identified by the author a long time ago [4][8][9] and many other researchers, is starting to be overcome, allowing the improvement of prediction methods and measurement capabilities.

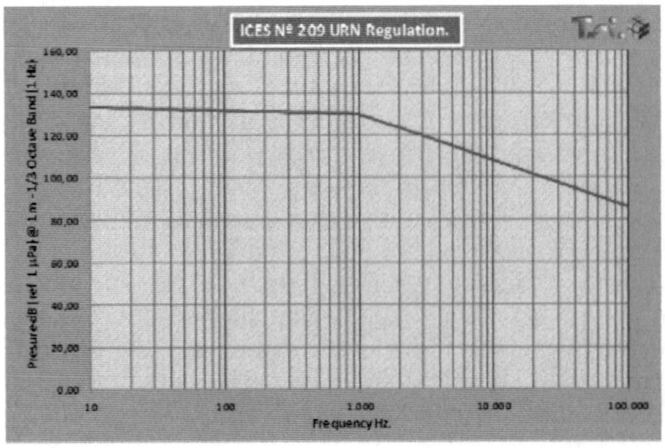

Figure 1: ICES N° 209 Regulation. URN target limits for FRVs (Source: TSI)

Taking into account concerns about harmonization and transparency, the author wants to warn that there are some ships of recent construction that are wrongly classified as "Oceanographic ship" because they really are not. A set of conceptual mistakes has been identified in their design, the inadequate selection of their main machinery or its assembly on the structure, making it impossible for them to comply with the URN limits of the ICES in the low-frequency range (10 Hz to 1 kHz). For the transparency of the market, it is essential that all FRVs should have a corresponding certificate issued by an independent company which guarantees that it strictly complies with the criteria defined by ICES. As an example of this certificate, Figure 2 below shows a URN signature card of FRV Ramón Margalef, showing the thresholds and the URN emitted in all frequencies.

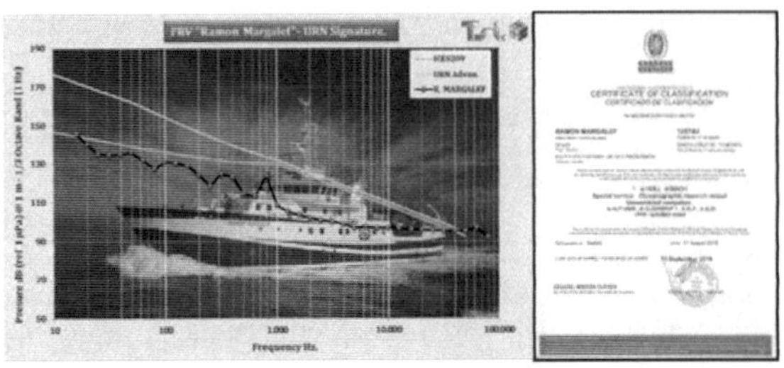

In January 2010 DNV published a Silent Class Notation as a new standard for the measurement of the underwater noise emitted by ships [10]. This standard specifies targeted limits and associated measurement procedures for shallow waters.

Depending of the compliance with the targeted limits the Classification Society provides different type of Notations: "A", "S", "F" and "SILENT-R", being the last one for vessels engaged in research or other noise-critical operations. The URN limits for this last category "R" are the same required by ICES CRR N°209.

Regarding to the measurement procedure the following points are highlighted: (i) only one hydrophone is employed (see the Figure 3); (ii) according to this Class Notation, DNV-GL decides the speed/s at which the vessel shall be tested, based on the vessel operational speed profile for typical hydro-acoustics operations preliminary submitted to it; (iii) only two runs (one for each side of the vessels) are required; (iv) TL parameter (Transmission Loss) is calculated according to the following formula: TL= 18 Log_{10} DCPA (DCPA stands for Distance of Closest Point of Approach) and a correction of minus 5dB must be included due to surface reflections; and finally (v) another important issue is the "data window length" used.

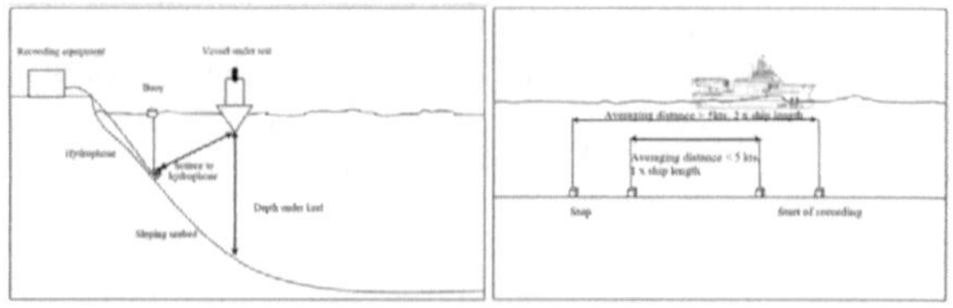

Figure 3: DNV-GL SILENT URN Measurement Procedure (Source: DNV-GL)

Regarding (ii), it is important to keep in mind that the speed of the ship becomes essential for the suitable dynamic-acoustic design of the silent vessels, and it should be defined and agreed at the beginning of the project and the Classification Society should not decide on it or change it at their convenience. For instance, ICES CRR209 defines its contractual noise limits at the speed of 11 knots, driving the entire dynamic-acoustic design of the ship to comply with this operational condition; this is established at the signing of the contract.

Additionally, this Classification Society proposed an alternative way to obtain noise radiated by ships through making a hole in the aft part of the hull of the ship, close to the propeller, to measure noise with a hydrophone or a pressure sensor. Despite this proposal being presented and discussed in technical meetings of experts promoted by European shipyards as an alternative to reduce time and cost of URN measurement, it was rejected by the majority of attendees.

1.2 (e) BV-NR 614 URN Notation

In October 2014 Bureau Veritas (BV) issued its first URN Notation, BV NR614 [11] along with a proposal of a European URN measurement standard [12] and [13] in the framework of AQUO project after a collaboration with TSI S.L. in this R&D project [1]. AQUO project was focused on underwater noise emitted by vessels, and its main objectives were to develop guidelines and recommendations to reduce shipping noise to achieve a quieter ocean and to provide support to policy makers, as well as to propose design solutions and mitigation measures to manage URN emission of ships to reduce its impact on the marine fauna. BV NR614 URN Notation was developed considering the URN database managed by TSI in SILENV project and AQUO project, bringing realistic, consistent, and reasonable justification of the down-selected limits and measurement procedure.

This notation aims to control and limit the environmental impact on marine fauna of all self-propelled vessels providing a URN measurement and compliance assessment procedure according to its specific URN vessel requirements, considering a minimum set of three runs for each side of the vessel and a set of speeds of the vessel according to its own operational modes. For further details regarding measurement and post-processing procedure, see the reference [11].

While ANSI/ASA is a URN measurement procedure that is focused mainly on deep waters (with limitation of its application in some areas of Europe) and DNV-SILENT measurement procedure is for shallow waters, BV NR 614 is the only URN measurement procedure that considers both deep waters and shallow waters, covering the widest range of measurements without any restriction.

1.2 (f) Other URN Notation of the Classification Societies: RINA, LR and ABS

The rest of the Classification Societies (CCSS): RINA (Registro Italiano Navale), LR (Lloyd´s Register of Shipping) and ABS (American Bureau of Shipping), released recently their corresponding URN voluntary Class Notations: Rina Dolphin on 1 July 2014, LR in February 2018, and ABS in July 2018. Most of these class notations have adopted the corresponding URN measurements procedures defined recently by ISO-17208-1 and ANSI/ASA.

1.2 (g) Measurement methodologies comparison and the importance of consistent URN data

Comparative analysis between the measurement procedures of the different CCSS reveals that all of them, with the exception of DNV-GL, are moving towards harmonization, adopting the same approach for TL (Transmission Loss), $TL=20\ Log_{10}\ DCPA$ recommended by ANSI/ASA and ISO-17208-1. It is rather curious that DNV-GL, without any technical or experimentally supported data, has adopted the formula $TL= 18\ Log_{10}\ DCPA$ instead of the one used by the others. A possible reason is that for a value of the DCPA around 100m, measurement distance range recommended by the Standards, DNV-GL, will always obtain a value for the underwater acoustic signature characterization of the ship (as a spherical source), 4dB below with respect to the values provided by the others procedures for the same ship. And that figure is without taking into account the additional reduction of 5dB justified as a correction due to the reflection of the noise on the seabed.

Accordingly, the same ship would present a completely different picture about its URN signature depending on how it is measured and how the data are analyzed. The repeatability of the test and comparability of results, which is one of the main objectives of the Standards, is definitively impractical and unfeasible. To support this controversial statement and with the sole purpose of contributing to the solution of this black spot of measurement heterogeneity, Figure 4 shows the underwater acoustic signatures of an FRV obtained with different post-processing methodologies. The input data have been the same in all cases and, as has been shown, the results vary depending on the methodology.

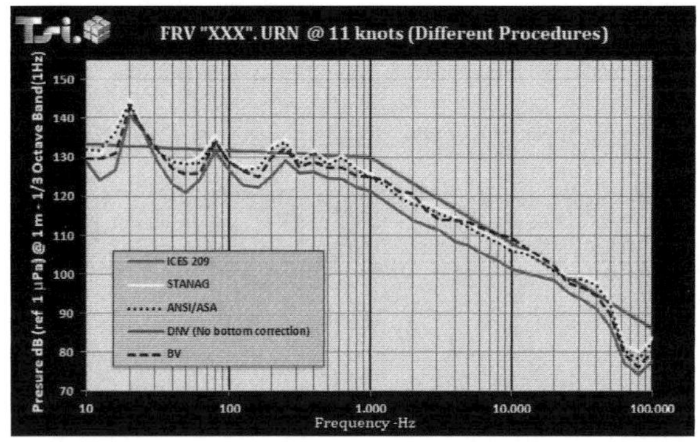

Figure 4: Comparison of URN results depending on post-processing methodology (Source: TSI)

It is clear that the availability of consistent experimental URN data is of critical importance as a starting point in the challenge of reducing a ship's URN. Fortunately, this issue is more and more consolidated, considering the movement of the marine sector and the different publications released lately related to URN experimental data analysis [14], and [15] as well as ECHO programme (Port of Vancouver, Canada) and R&D projects [1] such as AQUO, BESST, SILENV and SONIC, in which URN measurements of different vessels were conducted making possible the extension of the vessel URN database. Recently, it can be seen in the marine sector that some shipowners and equipment suppliers voluntarily demand URN measurements to better prepare themselves for future environmental limits.

1.3 URN SIGNATURE OF A SHIP DEPENDING ON THE CLASSIFICATION SOCIETY SELECTED: CONSENSUS AS A NEED OF THE SECTOR

It has been highlighted above that an agreement is needed to reduce the market confusion regarding URN emitted by vessels, to be efficient in the common commitment of its reduction. The risk of confusing the market is evident, and it should be solved by all the stakeholders, especially by the Classification Societies as independent parties of the marine sector. It is confusing and nonsensical that the same type of ship can receive the certification of silent vessel or not depending on the Classification Society chosen, especially where such methodologies or limits of acceptance have been defined without any technical support to endorse them. These issues, sooner than later, will be seen by a marine sector and a society worried about the pollution of the oceans, as a commercial strategy instead of a serious commitment to the challenge of reducing URN.

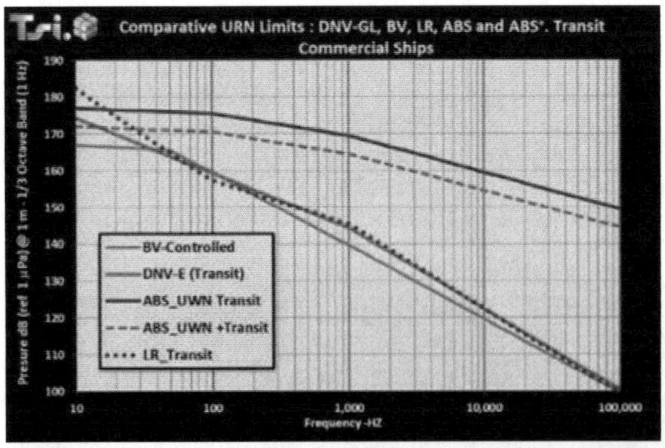

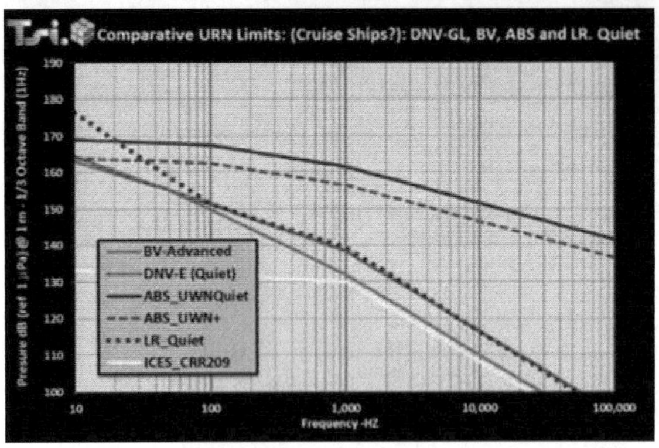

One of the main Classification Societies has released to the market a set of very flexible noise limits in all the frequency ranges where the noise induced by the machinery and propeller is always located, even when the propeller is cavitating. Once again, it will be very easy to understand what will be the most convenient choice from shipowners' and shipyards' point of view to avoid problems during the building and the obtaining of certificates.

As a comparative analysis, Figure 5 presents the URN limits required by several Classification Societies: BV, DNV-GL, LR, and ABS, for "Transit" and "Quiet" operational conditions, and for two types of ships, commercial vessels and cruise ships. The curve of limits required by ICES CRR209 for the FRVs has also been included. This type of vessel is named as "R" by all the Classification Societies that maintain for them the same criteria adopted in 1995 by ICES CRR209.

As it can be seen in Figure 5-a), while the limits required by BV, LR, and DNV-GL for the same commercial ships in Transit condition are very close for the selected frequency range (10Hz-1kHz), the ABS's limits are much less restrictive than the others, with deviations that vary from +18dB (at 100Hz) and +25dB (at 1kHz) to +40dB (at 10kHz). In the case of cruise ships or commercial ships in Quiet condition the comparative trend is almost the same. BV, LR, and DNV-GL require noise limits very close while ABS's limits are above those. In this case the deviations are around: +16dB (at 100Hz), +22dB (at 1kHz) and + 41dB (at 10kHz).

Secondly, in Figure 5-b), for comparative analysis the noise limits curve required by ICES CRR209 for the FRVs has also been included. As can be seen, the noise limits required for any FRV are located, at least, -36 dB, in the frequency range 10Hz-1kHz, and - 43 dB, at 10kHz, below the noise limits required by the least restrictive Classification Society, for any cruise ship or commercial ship in Quiet condition. It should be reminded that FRV sector is a small segment of civil ships that must respond, as it is already doing, to the most restrictive limits of URN. The marine sector must find the right guidelines to reduce the URN by all types of ships through technical advances and innovations applied to FRV technology in the last two decades.

Curiously, all classification societies have achieved a harmonization agreement for the FRV not only with regard to noise limits in all frequency ranges (10-100kHz), but also in relation to the operational speed conditions under which the experimental tests for assessment must be performed. Thus, the dynamic and acoustical design of these ships should be performed to meet the noise limits at 11 knots. This is well known since the beginning of each project. On the contrary, the ambiguity of the term Transit and Quiet from some of the URN notations, regarding the speed (power rate) in which experimental tests are performed, should be clarified from the engineering point of view.

As a summary, while the FRVs as civil ships must face the strictest URN limits at the defined speed of 11knots (also adopted by all Classification Societies), the rest of commercial vessels are able to select the laxest Classification Society as the most convenient one from which to obtain certification. Another inconsistency is related to the operational speed of the URN measurements for commercial vessels, since it is not defined (unlike FRV sector) or is a decision of the Classification Society, creating possible suspicions and mistrust about the methodology followed.

In order to clarify this abovementioned black spot for all ships that have obtained the certification of silent ship a set of parameters should be publicly accessible to avoid the lack of transparency in the marine market: corresponding ID card (underwater acoustic signature) of the vessel, speed tests, and information related to the measurement procedure used. At the very least, this clarification should be addressed for ships which have access to economic incentives coming from public money for the prevention of fraud or abuse.

1.4 TECHNICAL TOPICS RELATED TO THE MAIN URN SOURCES

There is some light at the end of the tunnel. The good news is that a harmonized format has been developed to present URN spectra adopted for most of the professionals of the URN sector, presenting the pressure (ref 1 μPa) at 1m of the hull in 1/3 Octave Band (1 Hz) against frequency (Hz) in the range from 10Hz to 100kHz. Figure 4 shows an example of this representation of the URN signature.

In parallel, from long ago, an intense debate has raged about which were the main sources that contribute to URN in a vessel. Traditionally, the propeller was considered as the principal root cause of the URN signature of the vessels, without any experimental support. These kinds of assertions were made based on outdated information, in which the propeller appeared as one of the main root causes of vibrations and noise on board. To bring light into this matter, in the framework of the SILENV project, a set of on-site measurements activities were carried out to obtain the URN signature with satisfactory frequency accuracy and resolution (0.5Hz) in order to identify in the energy spectrum the contribution of each one of the main URN sources (machinery and propeller). Figure 6 presents the spectrum in narrow band corresponding to a ROPAX ship as an example of the results obtained. No propeller cavitation was measured in the URN signature.

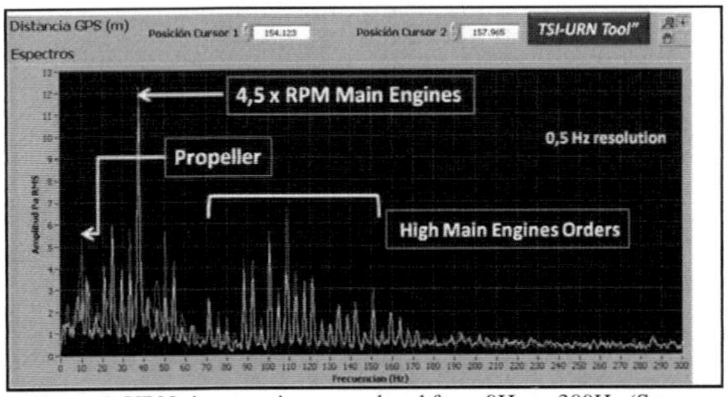

Figure 6: URN signature in narrow band from 0Hz to 300Hz (Source: TSI)

Figure 6 shows that the narrow band spectrum analysis allows with enough accuracy the identification of the contribution of each main source, both machinery and propeller, and, depending of each case, the global noise radiated by each ship. This sort of analysis demonstrates that the other sources must be considered for URN and neglecting the traditional assumption of the propeller as the unique source.

According to this argument, the abatement of the URN signature of new vessels requires a preventive control of the noise sources as well as the application of actions focused on reducing the vibrational energy transmission from the main mechanical sources, such as the main and auxiliary machinery, to the hull. Likewise, the intensity of the excitation induced by the propeller must be controlled as well as its cavitation.

To tackle both issues, propeller and machinery, new guidelines provided recently by IMO are in line with the tools, work, and studies carried out by the author and his company (TSI) throughout his career [16][17][18][19][20]. IMO recommends the use of hydrodynamic simulations based on CFDs, FE models, and the suitable combination of BEM and SEA models.

2. TOWARDS HARMONISED STRATEGY OF URN REDUCTION OF SHIPS

Shipping noise and its impact on marine fauna and environment is a global issue that cannot be reduced by acting only in one or another maritime region. Thus it requires global and harmonized efforts by all stakeholders and countries around the world. In this sense, Canada's initiative deserves especial mention. It is the results of the efforts made by Vancouver Fraser Port Authority who in 2014 started the ECHO (Establishing Cetacean Habitat and Observation) programme to better understand and manage the impacts of shipping noise in the endangered resident population of Southern Resident Killer Whales (SRKW) located in their natural habitat in the Salish Sea (British Columbia). In a clear vision of the maritime market, the policy applied by Vancouver Port consisted in the reduction of the port's taxes for all ships that could prove that they are implementing mitigation measures or changing their operational profiles to reduce their URN emissions.

Once again, as public money is involved, the transparency and homogenization of measurement procedures and limit values are essential. The next paragraphs summarize a set of proposals addressing several corrective actions that would contribute to clarify the current confused scenario:

- **Harmonization of the Regulatory Framework by the different States:** Even though it is not an easy task for the policy-makers, the regulatory framework should be harmonized, applying different mechanisms such as specific requirements stated by IMO as well as the different states, ports, and flag administrations.
- **Harmonization of the URN Measurement Procedures:** Significant efforts have been carried out by ANSI/ASA and ISO in this sense, particularly related to the URN measurement procedure in deep waters. But due to water depth limitations observed in the proximity of several very active shipbuilding areas, additional efforts should be applied by both regulatory bodies to the measurement procedure in shallow waters. Here the author suggests it is useful to review the "European URN Measurement Procedure" for both, "deep" and "shallow" waters, and its corresponding uncertainties assessment. The homogenization is also strongly required in terms of propagation (transmission loss) model and seabed effects model for comparative analysis and practical engineering applications. These topics were addressed, during three years of research in the framework of the AQUO project, by a wide variety of experts including shipyards, Classification Societies, universities, research centers, towing tanks, marine biologists, and engineering consultants.
- **Harmonization of the URN Limits of the different Classification Societies (CCSS):** The harmonization of URN limits for the different types of ships is an absolute "must" in order to avoid acoustic signature discrepancies for the same ship depending on the classification society chosen, that might make irrelevant the labeling of a vessel as silent or noisy. A desire for harmonization is a permanent clamor in all

technical meetings promoted by FINCANTIERI (one of the most active worldwide shipyards specializing in the construction of cruise vessels), and with the participation of all relevant Classification Societies and engineering experts. As a result of these technical meetings, BV and RINA are moving forward to the harmonization of their corresponding URN limits and corrections. On the other hand, the dual role assumed by DNV-GL, acting as URN consultant and also as a certification entity with its own URN limits, does not seem to be compatible and consistent with the transparency demands of the market.

- **Management of URN (old fleet and new ships) and URN Identification Card:** It is challenging to control and manage the environmental impact of a huge fleet of existing ships with an average age of more than twenty years. Conservative actions might be carried out to reduce the URN of the fleet focusing on operational mitigation measures such as navigating far away from protected areas and reducing their speeds. In both cases, "how far away" and "how far to reduce the speed" are technically undefined. These types of actions, which have significant impact on the exploitation of the ships, will drive shipowners to include technical changes in their future vessels to reach future limits of URN. In the meantime, the required distance and speed should be defined based on real underwater signature of each ship, by means of an URN Identification Card (URN_ID) that would include the URN levels measured at different operational speeds by certified independent companies. Additionally, for the FRV sector, the URN_ID would avoid the wrong denomination of "Fishery and Oceanographic Research Vessel" to vessels that do not meet the URN limits required by ICES CRR209.

3. METHODOLOGY PROPOSAL: NOISE AND VIBRATION COMPREHENSIVE MANAGEMENT (N&V-CM).

As has been mentioned previously this methodology is based on the experience of the author and his company solving noise and vibration issues in the marine sector and other industries for more than forty years. Its validation is based on successful results provided by the application of the Noise & Vibration Comprehensive Management (N&V-CM) methodology of different types of vessels such as luxury yachts, passenger ships, ferries, naval ships and all kinds of commercial ships with the strictest onboard N&V requirements.

Confirmation has been obtained in the last two decades after the dedicated application of this methodology and the experimental results obtained in the dynamic and acoustical design of seven research vessels, with not only high onboard N&V requirements but also with the compliance with the first and unique URN requirements for research vessels (ICES 209, see section 1.2 (c)).

From this author's point of view, for a positive result the main pillar of this methodology is the correct application of all the preventive Noise & Vibration control techniques from the earliest stage of the project (conceptual phase) to the final stage (building of the ship and sea trials). Figure 7 shows the main stages considered in this validated methodology for new vessel constructions.

Figure 7: Stages of N&V Comprehensive Management methodology for new vessels construction (Source: TSI)

3.1 BASIC PRINCIPLES OF THE METHODOLOGY

Vessels are elastic systems under periodic forces that come from different sources. Consequently, they are vibration sensitive. The vibration levels obtained in the system (the vessel) depends, mainly, on three parameters: (i) the intensity or magnitude of the excitation forces; (ii) how stiff is the structure; and (iii) the dynamic amplification at different frequencies due to resonance phenomena, local and global. As a consequence, the possible actions that could be implemented if we want to keep vibration levels under certain limits are the following: (A) minimize the excitation forces of the system; (B) avoid flexible structures from a dynamic point of view; and (C) avoid resonance phenomena (coincidence of structural frequencies and excitation frequencies).

Similarly, and from an acoustic point of view, the vessel has built-in sonorous sources: main engine and auxiliary ones, propellers, hydraulic systems, HVAC, etc., that are airborne and structure borne noise generators. The noise is transmitted or spread through all the structure of the vessel (path) and reaches all the different locals and spaces (receptors) as well being radiated to the water. In the same way, the possible actions that could be implemented to maintain the noise levels under the predefined limits are the following: (i) minimize the sound power and vibration power at the noise sources; (ii) reduce or diminish their transmission to the paths; and (iii) isolate the receivers.

From an elastic-acoustic point of view, the hull becomes a wave radiator to the water due to the acceleration induced on it by the action of the main mechanical excitation sources and to the propeller. In this case, preventive actions and sources control shall be applied focused on reducing the vibrational energy transmitted to the hull by the main machinery and by the propeller. So, dedicated FATs (Factory Acceptance Tests) should be done for the main machinery, and pressure pulses induced by the propeller and cavitation tests should be carried out to avoid this undesirable phenomenon.

As a summary it could be stated and assumed by all the parties involved - designer, shipyard, main supplier, insulation supplier, etc. - that, taking into account the strict N&V requirement defined in the contract specifications, the design principle that the "Dynamic-Acoustic Design of the vessel" should be mandatory at all stages of the project. The role of the consultant, recommended to be involved in this type of project, is to verify and guarantee that all parties do their part.

3.2 PRACTICAL APPLICATION OF THE METHODOLOGY

For the practical application of the previous "Basic Principles" the N&V Control activities are subdivided into two main groups. In a first module, within the Noise and Vibration Comprehensive Management (N&VCM) and as a control mechanism, it is critical and essential to incorporate in the purchase specifications of the different supplies the "dynamic and acoustic specific requirements" of all of the main N&V sources as well as the corresponding procedures to the Factory Acceptance Tests (FATs) that will be performed to control and verify compliance with those requirements before installing the most critical supplies on board. This first activity level has focused on the fact that the shipyard can develop a control of the main N&V and URN sources that are usually under the scope of the suppliers. The intention with this control mechanism is to achieve and guarantee the so-called "contractual sensitivity" of the suppliers towards the dynamic and acoustic objectives of the project.

If this task is not performed properly at the purchase stage, the shipyard must aware that the project could fail or, in many situations, the solutions or correction countermeasures that must be implemented will be more expensive.

The shipyard must assume, from the beginning of the project, the fact that the "Dynamic and Acoustic Vessel Design" is mandatory in all the design and building stages. Accordingly, once the N&V and URN main sources are under control by the clear definition of the dynamic and acoustical requirements contractually required of them, a second module or block of control activities must be applied. Included in this second group of activities are those that are within the sole competence of the shipyard, such as the supply of the vessel structure with an appropriate dynamic-acoustic design, without global and local resonances and with enough and suitable acoustic isolation to guarantee compliance with the N&V and URN contract requirements.

Designed to achieve the contractual requirements related to the compliance with the limits of noise and vibration on board as well as the URN limit, signed between the owner and the shipyard; the shipyard, within the second group of control activities, must apply with the assistance of its N&V consultant a whole and dedicated package of simulation tools and dynamic tests, such as N&V prediction calculations, and Radiated Noise, outdoor and underwater, prediction calculations, on achieving the optimal design of the ship.

Bearing in mind the two previous groups of control activities, they are critical and essential to guarantee the design and build of any kind of "silent ship". The following paragraphs will briefly describe several aspects that should be considered according to the Figure 7 scheme.

3.2 (a) Project definition stage: Vessel Specification and Preliminary Assessment of the Design

At this stage the suitable definition of the vessel specification, for which the participation of one expert N&V consultant is recommended, and the preliminary assessment of the conceptual design are essential for the success of the project. An assessment from the N&V point of view must be carried out in the specification of the vessel project, the structural design, noise insulation, and suitability of specified machinery, among other parts of the contract. It is absolutely mandatory that requirements regarding URN and onboard Noise & Vibration for this new vessel are defined in the vessel specification and a vague reference to ISO or IMO rules is not recommended. These requirements should be based on results from the preliminary assessment of the existing fleet, the vessel dynamic characterization including onboard noise and vibration prediction and measurement, existing and expected future regulation, comfort criteria, classification society, technological state of the art, N&V limits of the outfitting for its purchasing, and cost-benefit analysis of the selected choices with the possibility to include URN measurement to verify the compliance of the vessel with the selected standard and/or URN prediction. In this regard, these requirements will be part of the contractual conditions of the new vessel construction.

Therefore, this stage includes the analysis of the design criteria of the main N&V Sources and the establishment of contractual clauses in their purchase specifications. Considering that the main and auxiliary machinery along with the propeller/s are the main N&V sources, the suitable choice of them and the definition of specific dynamic and acoustical contractual criteria for these components become an essential and critical aspect to guarantee the compliance of the resulting vessel with the N&V requirements stated by the shipowner specification. In the end, the resulting level of vibration depends on the equipment itself as well as on the solution adopted for installation along with the interaction of installed components.

Thus, for the generation of reliable prediction models to simulate the dynamic and acoustical behaviour of the new ship, it is essential to have a reliable and experimental dynamic and acoustic characterization of the different main sources of noise and vibrations of the ship. This information must be committed and delivered by the different suppliers at the signature of the purchase contracts. The absence of this information, from the author's point of view, must be cause for rejection of the supply, given the high risks of lack of compliance as well as an increase of costs for the shipyard. These models are per-formed in two steps:

- **Conceptual design: Preliminary Assessment.** The objective of this step is to make sure that the new vessel is designed right from the beginning with URN and onboard N&V in mind. Decision-making within this phase has a huge impact on the design, construction, and operational complexity and costs. The underestimation of attention to detail during this step could lead to a situation in which the project will become too expensive or unfeasible to achieve the established URN and onboard N&V goals. The complexity of this step is relatively low and it relies highly on the N&V performed analysis, which is related to the general distribution of the vessel, its structural continuity and conceptual selection of the main machinery and their mountings, and its auxiliary equipment that has a significant emission of noise or vibrations. These recommendations should be considered in the ripening process of the vessel GA (General Arrangement) and preliminary structure drawings.
- **Preliminary approach to the acoustical behaviour of the new ship.** Once the structure of the vessel is defined, and the dynamic and acoustical characterization of the main sources of noise and vibration are determined, it is feasible to get, based on experience, a preliminary approach to the expected acoustical behavior of the new ship, identifying the weaknesses in terms of relative location between sources and receivers as well as the noise insulation material that will be necessary to get the compliance with the goals. The results of these calculations are used as: (i) input for changes in structure, distribution of the vessel, or even in type of equipment to be installed if it is necessary; (ii) input for optimization of the insulation installation; and, (iii) as baseline for following stages of the project until the vessel sea trials.

3.2 (b) Purchasing stage: Awareness of Suppliers

As has been mentioned previously, this stage focuses on the sensitization of equipment suppliers to the amount of noise and vibration levels allowed, and the shipyard´s request for a quotation of the necessary equipment considering these levels. Suppliers' contribution to the full success of the vessel project is beyond dispute. Their equipment, i.e. main and auxiliary machinery, propeller, thrusters, HVAC, accommodation panels, insulations, etc., must meet with the contractual requirements in terms of vibration and noise to be purchased by the shipyard. Suppliers must be able to provide N&V data and spectra to justify fulfilling the limits requested in the contract. This stage starts before the building stage to consider lead times. Nevertheless, both stages sometimes overlap.

Furthermore, the information provided by equipment suppliers is also used to improve the prediction models of onboard N&V and URN performed in the next stage.

3.2 (c) Building stage: Simulation, control and corrective actions

- **Simulations:** Obtain the expected vibration and noise levels in all the spaces of the ship as well as the expected URN of the ship (see Figure 8) by the application of the most recent state-of-the-art simulation software based on FEA, BEM, and SEA models (depending on the frequency band). The comparison of these expected levels with the limits required for each parameter will enable the application at minimum cost and, in each case, the corresponding corrective solution to achieve the desired goals.

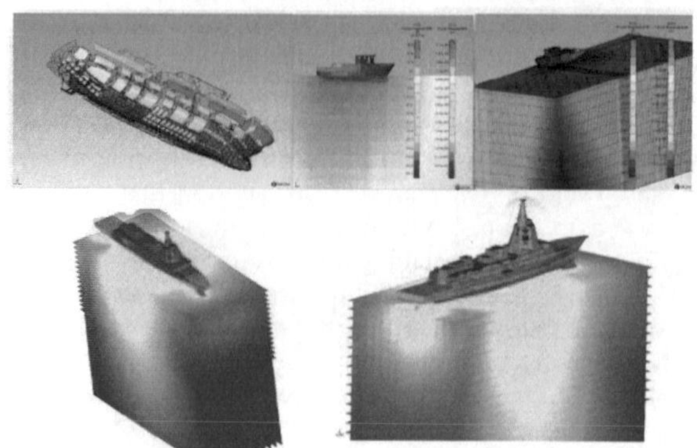

Figure 8: URN prediction through FEM-BEM-SEA model (Source: TSI)

- **Factory Acceptance Tests (FAT) and Site Acceptance Tests (SAT):** The objective is the verification of compliance of the supplied equipment with the contractual requirements defined in the project and purchase stages as well as being able to perform corrective actions in early stages. If this step is omitted, and in the case that the equipment to be installed does not meet the specifications, the final result will not correspond with predictions.
- **Inertia and mobility tests of machinery foundation:** These tests are performed to ensure proper installation, and to guarantee the absence of resonance phenomena in the foundation structure of the main and auxiliary engines once this equipment is integrated into the vessel structure (see Figure 9).
- **Airborne sound insulation index measurement:** A set of tests should be carried out as soon as the

accommodation begins to be installed to carry out corrective actions, in case of non-compliance results with regards to specification.

- **On-site survey during accommodation and noise insulation assembling process:** The achievement of the predefined target of noise on board is related not only to the suitable selection of noise insulation materials (according to the technical properties used in the simulations) but also to the right assembly of these components.

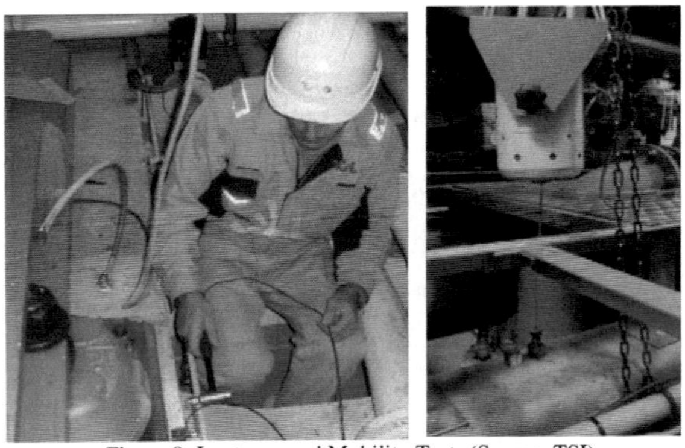

Figure 9: Inertance and Mobility Tests (Source: TSI)

3.2 (d) Sea trials stage: verification of onboard N&V and URN

At this final phase of the project, a set of measurements is performed to validate and verify that contract requirements aa stated by the Classification Society have been successfully met. The sea trials usually include onboard vibrations and noise, but barely airborne radiated noise is not included in URN measurements (unless it is eventually incorporated in the vessel specification) because there is no compulsory regulation in this regard.

As can be seen in Figure 10, and contrary to other consultant competitors, including the DNV-GL that act as consultant and certification agencies, it is an habitual practice of TSI to present the results of the URN measurements of the different vessels in which they have acted as N&V consultant. They correspond to the FRV´s "Ramón Margalef" and "Sarmiento de Gamboa" that have been measured according to the procedure defined by BV in the NR 614 URN Notation.

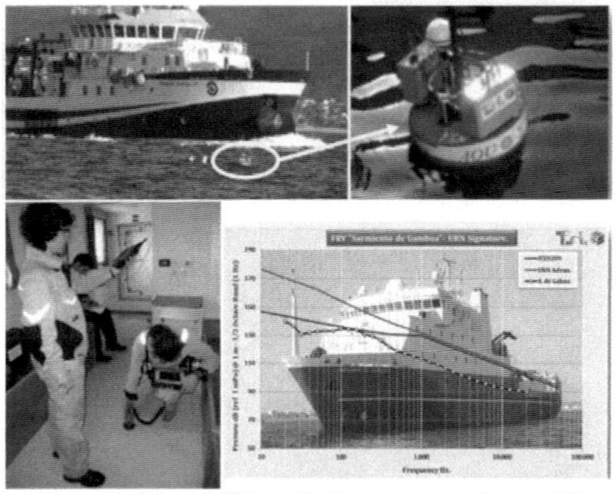

Figure 10: Onboard N&V and URN measurements (Source: TSI)

From the author's point of view, these measurements, especially URN measurements, might set a performance baseline of the vessel, which can be used for future condition monitoring of the vessel, helping ship operators to find any anomaly or malfunction in the vessel.

In parallel, these sea trials are also an extraordinary opportunity to perform additional measurements for R&D purposes or improvement of future vessels designs and shipbuilding.

4. CONCLUSIONS

Based on the previous information, the following conclusion could be pointed out:

- TSI's methodology (N&V Comprehensive Management) has proved its capability to achieve quieter vessels according to its portfolio and successful references.
- The upcoming URN regulatory framework will force profound changes in the whole marine industry in terms of design, building, and operation. It is preferable to be ready and proactive considering these changes.
- Research vessels sector is a small segment of civil ships that must fulfil the most restrictive limits of URN. This is accepted and there is consensus among all Classification Societies. Consequently, research vessels have become a technological reference and a clear example of the technical advances and innovations that can be applied to the rest of the civil shipbuilding industry to minimize the vessel impact in the oceans. Of course, other factors such as cost/benefit ratios and bunkering consumption efficiency should be considered and properly weighted.
- The validation of a vessel as a research vessel, and thus, as a technological reference, should be performed by an independent third party, issuing the corresponding official certification including a possible vessel URN_ID to control its noise emissions.

- The harmonization of the URN measurement procedures and URN limits should be mandatory to avoid confusion in the marine sector and to increase the competitiveness of all the stakeholders involved in this market. Only in such cases will it be possible to guarantee that effective innovations and technical solutions will be useful to other potential shipowners and will avoid replication of mistakes of research vessels that do not in fact fulfil the ICES CRR 209 limits.

ACKNOWLEDGMENT

The author wish especially thank to all the members of the company TSI (Técnicas y Servicios de Ingenería), which made this work possible thanks to TSI participation in European R&D projects related to URN (BESST, SILENV and AQUO) and their technical support through the onboard N&V and URN measurements as well as the numerical calculations carried out over the last years.

REFERENCES

1. BESST- Breakthrough in European Ships and Shipbuilding Technologies, SILENV- Ships oriented Innovative soLutions to rEduce Noise & Vibration and AQUO-Achievement Quiet Oceans, projects founded by the EC within FP7.
2. DIRECTIVE 2008/56/EC OF THE EUROPEAN PARLIAMENT AND OF THE COUNCIL of 17 June 2008 establishing a framework for community action in the field of marine environmental policy (MSFD).
3. IMO. Guidelines for the reduction of underwater noise from commercial shipping to address adverse impacts on marine life. MEPC.1/Circ.833 7 April 2014.
4. BELTRAN, P., et al, 'Achievement of the new underwater radiated noise requirements by the Spanish Shipbuilding Industry. The FRV Ramon Margalef'. Proceedings of the 11th European Conference on Underwater Acoustics.
5. ANSI/ASA S12.64-2009/Part 1: Quantities and Procedures for Description and Measurement of Underwater Sound from Ship – Part 1: General Requirements.
6. ISO-17208-1 :2016. Underwater acoustics - Quantities and procedures for description and measurement of underwater sound from ships- Part 1: Requirements for precision measurements in deep water used for comparison purposes.
7. ICES CRR Nº 209 Regulation.
8. BELTRAN, P., 'The Most Recent Noise & Vibration Assessment of the European Fleet, within the Framework of the SILENV Project'. Ship Science & Technology - Vol. 8 - n.° 14 - (43-66) January 2014 - Cartagena (Colombia).
9. BELTRAN. P., 'Reduction of Underwater-Radiated Noise from Ships: New Shipbuilding Challenge. The Vessels Ramón Margalef and Ángeles Alvariño as Technological References of How to Build Silent Vessels'. Ship Science & Technology - Vol. 8 - n.° 15 - (45-60) July 2014 - Cartagena (Colombia).
10. DNV-Silent Class Notation. Part 6, Chapter 21, January 2010.
11. Bureau Veritas, Rule Note NR614 DT R00E, Underwater Radiated Noise (URN). Bureau Veritas, Oct. 2014.
12. AQUO European R&D Project, deliverable D3.1, European URN Standard Measurement Method, April 2014.
13. MORENO, A., et al, 'New European Underwater Noise Measurement Procedure Standard Developed in AQUO project'. May 2015 IEEE Oceans15 Genoa.
14. MCKENNA, M.F., et al," Underwater radiated noise from modern commercial ships".
15. GASSMANN, M., et al. 'Deep-water measurements of container ship radiated noise signatures and directionality'.
16. BELTRÁN. P., 'Noise vibration on ship prediction: Basic engineering tool to fulfil the current requirements of comfort and quality'. Naval Engineering Magazine, 2001.
17. BELTRÁN. P., et al, 'Silent Fishing Vessels. A milestone for small and big Spanish shipyards. Lessons learnt'. Naval Engineering Magazine, 2006.
18. BELTRÁN, P., 'The Fishing Research Vessel Miguel Oliver'. Naval Engineering Magazine, 2007.
19. BELTRÁN, P., 'Oceanographic Ship Miguel Oliver: The excellence on Noise and Vibration on board fulfilling ICES-Nª209'. First Prize 47th Congress of Naval Engineering and Maritime Industry, 2008.
20. BELTRÁN, P., TESORERO, M.A. 'Spanish Shipbuilding ready to meet the new and demanding Environmental Requirements: New Challenges and Business Opportunities'. Ingeniería Naval #887. Silver Price 49th Congress of Naval Engineering and Maritime Industry.

AUTHOR BIOGRAPHY

Publio Beltrán is Managing Director at TSI SL (Técnicas y Servicios de Ingeniería) and a well-known naval architect and marine engineer. He is responsible for the management of TSI during the last forty years, carrying out projects of silent vessels and other noise and vibration analysis in several industrial sectors, among other engineering activities.

MECHANICAL CHARACTERIZATION OF POLYMER COMPOSITE MATERIALS FOR LONG LENGTH SHIPS

A.K. Haldar, T. Gobikannan, A. Portela, C. Athavale, and **A.J. Comer,** University of Limerick, Ireland

SUMMARY

In the marine industry Fibre-Reinforced Polymers (FRP) are currently dominating the manufacture of vessels up to 50m in length, with liquid resin infusion (LRI) being the most frequently used manufacturing technique, of which vacuum-assisted liquid resin infusion is the most widely adopted LRI variant. However, current regulations (SOLAS from IMO) restrict the use of composite materials in vessels over approximately 50m in length. FIBRESHIP is a Horizon 2020 funded EU project that aims to further the use of FRPs in vessels over 50m in length by addressing the regulatory restrictions and the numerous other challenges associated with manufacturing long-length FRP composite ships. The mechanical performance of new commercially available composite materials as potential candidates for selection in composite ship construction is central to this work. This paper provides an overview of selected work performed as part of the FIBRESHIP project in terms of evaluating various mechanical properties of selected laminates under dry and wet conditions. Test samples extracted from a glass fibre reinforced vinyl ester laminate were immersed in seawater at 35°C for durations of one month and three months. Three-point bend and interlaminar shear strength tests were performed to evaluate the retained flexural strength and interlaminar shear strength of the composite laminates post immersion. After three months of immersion the retained flexural strength and interlaminar shear strength was 90% and 79% respectively, highlighting the potential effect of seawater on the laminate and the importance of accounting for these effects during the design phase. Tested specimens were also examined using micro-computed tomography (μCT) to evaluate the failure morphology.

KEYWORDS

Flexural strength
Interlaminar Shear Strength
Glass Fibre
Vinyl ester

1. INTRODUCTION

Glass fiber based composite materials have been adopted for certain maritime applications for the last twenty-five years. Indeed, they were used in offshore oil platforms in the early 1990s and are also well known in the naval sector where they are often considered for the construction of vessels under 50m in length. There are also other quite common applications like water storage vessels, pipelines for water desalinization plants, and tidal current blade systems for energy generation [1].

In all these applications the corrosion resistance of glass fiber composite materials makes them very effective. Despite the benefits of using composite materials in naval ship structures, a long-standing concern is the loss of stiffness, strength, and other mechanical properties caused by sea water absorption. A potential problem with the absorption of water by composite materials (particularly as chemically-bound water) is the possibility of softening and damage to the polymer matrix and fibre-matrix interface [2, 3]. The effects of seawater are different depending on the type of resin used. There are several kinds of resin that could be used for maritime structures such as phenolic, polyurethane, polyester, vinylester, and epoxy resins of which polyester, vinylester, and epoxy are the most commonly used and are well studied in several scientific journal [4-5].

Absorbed seawater can degrade the strength and stiffness of composite materials particularly under tensile, flexural, compressive, and interlaminar shear loads [6-9]. Xu et al. analysed the residual compressive properties of E-glass/vinyl ester (VE) composites after exposure to sea water for twenty-nine months. The composite samples were then subjected to impact damage followed by residual compressive strength analysis, where one side of the composite sample was exposed to seawater. The results confirmed a 10% reduction in compression strength for seawater exposed impact damaged samples compared to the original dry samples [8]. Mittal et al. performed experimental studies to understand the influence of seawater on the mechanical and thermal behaviour of untreated and treated MMT (montmorillonite clay) based glass fiber/vinyl ester composites. The flexural strength of untreated and treated MMT composites was reduced by 7 and 4.5 %, respectively, compared to dry samples after immersion in seawater for 65 days [10].

However, limited work has been reported on the effect of seawater on the LEO fire retardant Vinyl ester resin from Buefa. Therefore, the aim of the current study is to evaluate the effect of seawater absorption on the mechanical properties (flexural strength and interlaminar shear strength) of laminates manufactured by vacuum assisted resin infusion using a glass fiber reinforced vinyl ester resin system.

2. MATERIALS AND METHODOLOGY

2.1 LAMINATE MANUFACTURING

Uni-directional (UD) glass laminates were manufactured by the vacuum-assisted liquid resin infusion process. The reinforcement fabric used in this study was Saertex U-E 940 g/m^2 LEO unidirectional non-crimp glass fabric. The reinforcement fabric used in this study has 90% of the glass fibres aligned with the 0° direction. The remaining glass fibres are oriented in the 90° direction to provide support to the dry reinforcement. The glass fabric layers were arranged in a $[0]_{2S}$ stacking arrangement. A vinyl ester resin (LEO Injection Resin 8500 from BÜFA) was infused into the fabric under ambient temperature and allowed to cure for 24 hours. A post cure of 6 hours at 80° was also performed. Both the fabric and resin are part of the LEO composite system [11]. Test coupons were extracted by rotating disc cutter for both flexural and interlaminar shear tests.

2.2. EXPERIMENTAL

2.2 (a) Flexure test

Three-point bend tests were performed under quasi-static loading conditions in accordance with ISO 14125. Samples were dried for four hours at 45 °C prior to testing. A nominal span length of 60mm was used at a testing speed of 1mm/min. The upper roller diameter was 10mm and the lower roller diameter was 4mm. Flexural strength, stiffness and strain in the longitudinal direction (0°) were reported. Apart from the dry state, samples were also immersed in seawater at 35 °C for durations of one month and three months. Flexural properties (strength and modulus) after immersion were also evaluated.

2.2 (b) Interlaminar shear test

Short-span three-point bend tests were conducted under quasi-static loading conditions in accordance with ISO 14130. A nominal span length of 15 mm was used, at a testing speed of 1mm/min. The upper roller diameter was 10mm and the lower roller diameter was 4mm. The apparent interlaminar shear strength (ILSS) in the longitudinal direction (0°) was investigated. Similar to the flexure testing, ILSS samples were also immersed in seawater at 35°C for durations of one month and three months. The inter-laminar shear strength after immersion was also evaluated.

2.2 (c) Micro X-ray computed tomography

Micro-CT scanning (Xradia versa XRM 500 Carl Zeiss X-ray Microscopy Inc.) was used to examine the failed flexural and interlaminar shear samples. Scans were carried out with an optical magnification of 0.39 x and an exposure time of 3.79 s. The x-ray source parameters were 80 kV and 88 μA with a pixel resolution of 10.7μm for all tomographic slices.

3. RESULTS AND DISCUSSION

3.1 EFFECT OF SEA WATER AGEING ON FLEXURAL PROPERTIES

Figure 1 shows the flexural strength values for dry and seawater aged laminates. The fibre volume fraction obtained based on thickness measurement in accordance with ISO 14127 are also shown. The dry sample exhibited the highest flexural strength (1100 MPa) and modulus (38.24GPa) compared to seawater aged laminates. The flexural strength of samples aged for one and three months decreased by 7% and 9% respectively. Strain at failure followed a similar trend to the flexural strength. It is difficult to draw a conclusion for the flexural modulus as the one-month sample showed a greater reduction compared to the three-month sample so further analysis is required of this aspect. Notably, the one-month sample had the lowest fibre volume fraction which could be a contributing factor to the relatively low modulus value observed for the one-month sample.

In the literature the decline in mechanical properties of hydrothermally aged composites has been attributed to the uneven thermal expansion of glass fiber and vinyl ester which caused swelling and hydrolysis at the fibre and matrix interface [12, 13]. An increase in the modulus of samples aged for three months was also observed in the literature and was attributed to matrix stiffening caused by the seawater ageing process [14]. Micro computed tomography (μCT) images of the flexural samples revealed the failure mechanisms as shown in Figure 1. Seawater aged samples mainly exhibited fibre buckling failure on the compression side as shown in Figure 1 (b) and (c).

Table 1. Flexural modulus and strain value of glass/vinyl ester laminates.

Sample	Flexural Strength (MPa)	Flexural Modulus (GPa)	Flexural Strain at Failure (%)	Fibre volume Fraction
Dry	1100.44±0.04	38.24±0.03	3.16±0.12	0.546±0.007
Sea water (1M)	1023.24±0.10	32.41±0.04	3.08±0.12	0.538±0.10
Sea water (3M)	1002.42±0.01	37.90±0.04	2.92±0.06	0.550±0.012

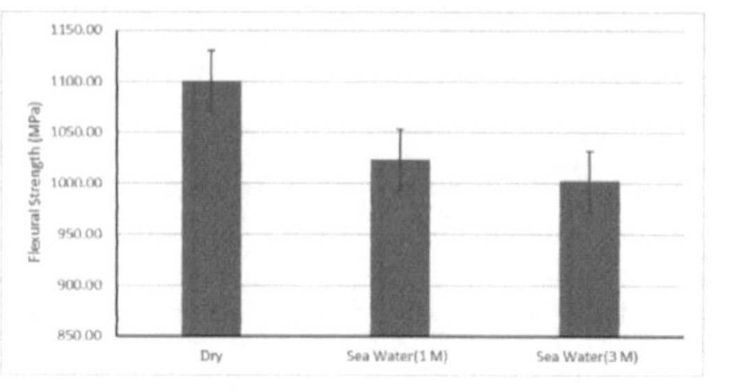

Figure 1: Flexural strength of glass/vinyl ester UD laminates as a function of exposure to seawater.

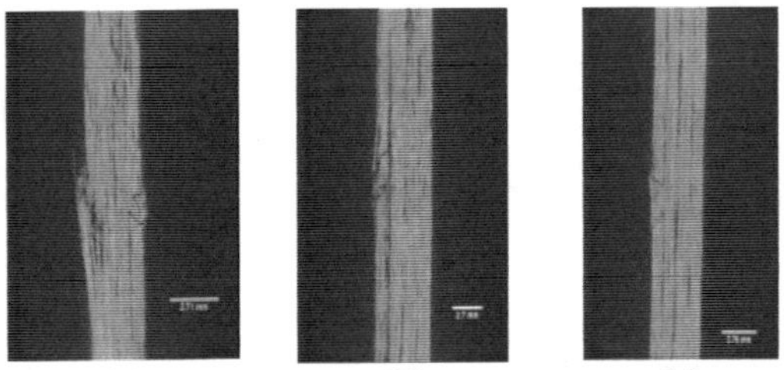

(a) (b) (c)

Figure 2: Micro-computed tomography images: (a) Dry; (b) 1 month; (c) 3 months, revealing failure mechanisms caused by the flexural testing.

3.2 EFFECT OF SEA WATER AGEING ON INTERLAMINAR SHEAR STRENGTH PROPERTIES

Figure 3 shows the interlaminar shear strength of glass/ vinyl ester laminates. The results follow a similar trend to the flexural strength behaviour reported earlier in this paper. The highest apparent interlaminar shear strength (55.30 MPa) was observed for the dry condition. Interlaminar shear strength decreased by 6% after one month followed by a reduction of 21% after three months. A reduction in the interlaminar shear strength is indicative of weakening of the fibre-matrix interface due to the absorbed water [14]. Micro-computed tomography images of the tested samples are shown in Figure 4. The expected damage mechanism is interlaminar shear cracking at mid-thickness but this is difficult to observe from the micro-CT images obtained. Further analysis is required using higher resolution micro-CT scans or SEM (scanning electron microscopy) to provide more detail regarding the failure mechanisms resulting from the interlaminar shear testing.

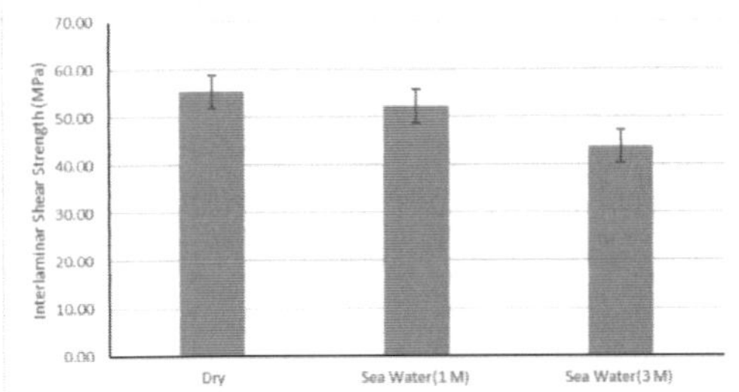

Figure 3: Interlaminar shear strength results for the glass fibre reinforced vinyl ester test samples.

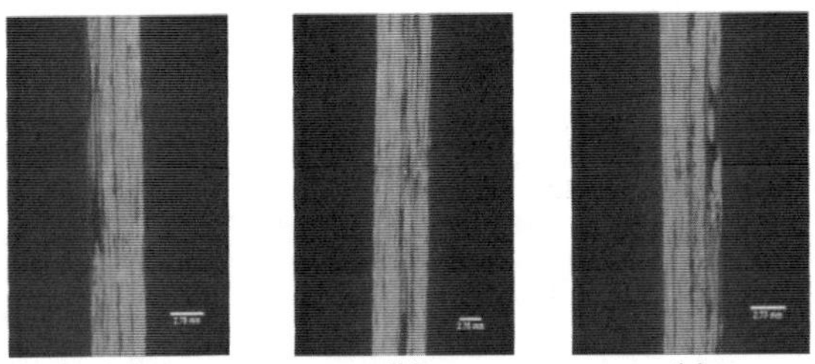

(a) (b) (c)

Figure 4: Micro-computed tomography images: (a) Dry (b) 1 month (c) 3 months.

4. CONCLUSION

Uni-directional glass fiber reinforced vinyl ester composite laminates were manufactured by the vacuum-assisted liquid resin infusion process. Test samples extracted from the laminates were subjected to three-point flexure and interlaminar shear testing to evaluate the retained mechanical properties after exposure to seawater for durations of one and three months. The highest flexural strength, modulus, and interlaminar shear strength was found for the dry laminates. Regarding the retention of flexural strength and interlaminar shear strength, the material retained over 90% of its flexural strength and 79% of its interlaminar shear strength after exposure to seawater for three months highlighting the influence of exposure to seawater on mechanical properties. Failure analysis was attempted using micro-computed tomography for both sample types. The technique was useful in highlighting micro-buckling in the flexure samples but was not successful in highlighting the damage occurring in the interlaminar shear samples at the resolution considered. It is intended to undertake further analysis of the failure mechanisms using either higher resolution micro CT or SEM. Overall, this study provides an insight into the effect of seawater immersion on selected mechanical properties of a glass fiber reinforced vinyl ester composite and the importance of accounting for these effects during the design phase.

5. ACKNOWLEDGEMENTS

The authors would like to acknowledge that this work is funded under the Horizon 2020 programme through the FIBRESHIP project (Project ref: 723360).

6. REFERENCES

1. B. AMELIE, D. PETER, T. FREDERIC, Sea water ageing of composites for ocean energy conversion systems: influence of glass fiber type on static behavior. *Appl Compos Mater*, 19, 459–73, 2012.
2. G. MAROM AND L.J. BROUTMAN, Moisture in Epoxy Resin Composites. *The Journal of Adhesion*, 12,153-164, 1981.
3. J. MIJOVIC, K.F. LIN, The effect of hygrothermal fatigue on physical/mechanical properties and morphology of neat epoxy resin and graphite/epoxy composite. *Journal of Applied Polymer Science*, 30, 2527-2549, 1985.
4. T. NGUYEN, K. AOUADI, D. ALSHEH, J. CHIN, Effects of civil engineering environments of interfacial properties of polymer/glass fiber composites. *International conference on composite engineering*, 1997.
5. E.P. GELLERT, D.M. TURLEY, Seawater immersion ageing of glass-fibre reinforced

polymer laminates for marine applications. *Compos A Appl Sci Manuf*, 30, 1259–65, 1999.

6. A.H.I. MOURAD, A.M. BECKRY MOHAMED, T. EL-MAADDAWY, Effect of seawater and warm environment on glass/epoxy and glass/polyurethane composites. *Applied Composite Materials*, 17, 557-573, 2010.

7. E.P. GELLERT, D.M. TURLEY, Seawater immersion ageing of glass-fibre reinforced polymer laminates for marine applications. *Composites Part A: Applied Science and Manufacturing,* 30, 1259-1265, 1999.

8. L.R. XU, A. KRISHNAN, H. NING, U. VAIDYA, A seawater tank approach to evaluate the dynamic failure and durability of E-glass/vinyl ester marine composites. *Composites Part B: Engineering,* 43, 2480-2486, 2012.

9. A. KOOTSOOKOS, A.P. MOURITZ, Seawater durability of glass and carbon polymer composites. *Composites Science and Technology,* 64, 1503-1511, 2004.

10. G. MITTAL, V. DHAND, K.Y. RHEE, S.J. PARK, H.J. KIM, D.H. JUNG, Investigation of seawater effects on the mechanical properties of untreated and treated MMT-based glass fiber/vinylester composites. *Ocean Engineering*, 108, 393-401, 2015.

11. N.H. NASH, A. PORTELA, C. BACHOUR, I. MANOLAKIS, A.J. COMER, Effect of Environmental Conditioning on the Properties of Thermosetting- and Thermoplastic-Matrix Composite Materials by Resin Infusion for Marine Applications, *Compos part B*, 177, 2019.

12. A. HODZIC, J.K. KIM, A.E. LOWE, Z.H. STACHURSKI, The effects of water aging on the interphase region and interlaminar fracture toughness in polymer–glass *composites, Compos. Sci. Technol*, 64, 2185–2195, 2004.

13. F. ELLYIN, R. MASER, Environmental effects on the mechanical properties of glass-fiber epoxy composite tubular specimens, *Compos. Sci. Technol,* 64, 1863–1874, 2004.

14. R.K. NAYAK, B.C. RAY, Influence of seawater absorption on retention of mechanical properties of nano-TiO_2 embedded glass fiber reinforced epoxy polymer matrix composites, *Archives of Civil and Mechanical Engineering*, 18, 1597-1607, 2018.

7. AUTHORS' BIOGRAPHY

Dr Amit Kumar Haldar is a postdoctoral researcher at the University of Limerick (ULim), Ireland. Dr Haldar has gained advanced knowledge and skills related to lightweight composite materials during his doctoral studies at the University of Liverpool and published four SCI journal papers in leading international journals, together with several conference papers. His recent research focuses on the mechanical properties of polymer composites manufactured by a variety of methods, including hot compression and liquid resin infusion.

Dr T Gobikannan is a post-doctoral researcher at the University of Limerick, Ireland. He is responsible for conducting research related to composite material production and characterisation. His previous experience includes teaching (Associate Professor Level) cum research positions in various technical universities for the past nine years, and assistant researcher position at Feng Chia University, Taiwan.

Dr Anthony Comer is a Lecturer at the University of Limerick (ULim), Ireland and is also a Principal Investigator at the Bernal Institute (Composite Materials Cluster) and the Irish Centre for Composites (IComp), also hosted at ULim. Dr Comer is the Principal Investigator of FIBRESHIP at ULim and has over fifteen peer-reviewed journal papers published in Q1 ranked journals. His recent research focuses on the mechanical properties and the effect of environment on polymer composites manufactured by a variety of methods including liquid resin infusion and automated tape placement.

FIRE SIMULATIONS OF A FISHING RESEARCH VESSEL WITH FRP STRUCTURES

A Tissari, T Korhonen, T Kling, A Korkealaakso and **T Hakkarainen,** VTT Technical Research Centre of Finland Ltd, Finland

SUMMARY

The fire safety effect of using fibre-reinforced polymer (FRP) as the primary construction material in a fishing research vessel was studied by fire simulations. The effect of FRP structures on fire development was assessed by comparing the simulated gas temperatures and potential heat releases with FRP and steel structures. The structural integrity of FRP structures was assessed using simulated temperatures of the structures as indicators of integrity. The effect of protective mineral wool and intumescent coating layers was also quantified. The results showed that despite the protection, the structural integrity of FRP bulkheads could be compromised in fire conditions. Mineral wool was found to be better protection than the intumescent coating: it can either prevent or postpone the pyrolysis of the FRP bulkhead, depending on the fire exposure.

NOMENCLATURE

CFD	Computational Fluid Dynamics
FDS	Fire Dynamics Simulator
FRP	Fibre-Reinforced Polymer
FRV	Fishing Research Vessel
HRR	Heat Release Rate
TGA	Thermogravimetric analysis

1. INTRODUCTION

The FIBRESHIP project is an ambitious innovation project to develop a new market focused on the construction of commercial vessels in composite materials (Fibre-Reinforced Polymers, FRP) greater than 50 m in length. The main objective of the FIBRESHIP project is to generate the regulatory framework that allows the designing and building of large-length ships in FRP material overcoming the technical challenges identified. In order to achieve this objective, the project is qualifying and auditing innovative FRP materials for marine applications, elaborating new designs and production guidelines, generating production and inspection methodologies, and developing numerical software tools capable of assessing the structural performance validated through experimental testing.

As a part of the work package devoted to design, engineering and development of guidelines, on-board fire events have been simulated to assess the fire safety effect of using FRP as the primary construction material of the vessel, compared to conventional steel structures. This paper describes the fire simulations of a fishing research vessel (FRV) with FRP structures. Several fire scenarios in an accommodation space of an FRV were simulated with a computational fluid dynamics (CFD) software Fire Dynamics Simulator (FDS) [1], version 6.7.1. To create different fire scenarios, the utilized design fires, bulkhead structures and ventilation conditions were varied in the study.

From fire safety aspect, the following matters were of primary interest:

- the effect of FRP structures on fire development in comparison to more conventional steel structures,
- the integrity of the FRP structures in fire conditions,
- the effect of protective material layers on the fire behaviour of the FRP structures.

The effect of FRP structures on fire development was assessed by comparing the gas temperatures and the potential heat releases obtained from simulations with FRP structures to reference simulations where all structures were conventional steel structures. The potential heat release was defined as the heat release due to the complete combustion of all produced gaseous fuel. Due to ventilation-controlled conditions in the studied enclosure, not all produced fuel gas burned. As the studied enclosure was located within a ship superstructure, the unburnt fuel could combust in an adjacent compartment and thus cause rapid fire spread.

As the structural responses of the FRP structures were not explicitly studied in this work, the simulated temperatures of the structures were used as indicators of integrity. It is known that after reaching the glass transition temperature the material will lose a significant portion of its load carrying capacity. The glass transition temperature of FRP materials is typically about 100°C. Due to the uncoupled nature of the simulations (the structural deformation is not taken into account), there is significant uncertainty in the results after the FRP structures reach high temperatures.

To be able to quantify the effect of protective material layers, two different protective layers on top of the FRP structures were used in the simulations. The increase in the potential heat release, the duration of fuel generation of FRP structures and the extent of pyrolysing areas were compared between the simulations

In addition, one objective of this study was to gain better understanding of simulating enclosure fires with FRP structures. It is to be noted that pyrolysis modelling of complex materials, simulating fires including structural responses or simulating ventilation-controlled enclosure fires, is not by any means mature. The results of this study are thus limited in application.

2. SIMULATION INPUT

2.1 GEOMETRY

The simulated space was an accommodation space designated as a multifunctional space in the ship's general arrangement. The longitudinal bulkhead on the starboard side is exterior. Visualization of the room geometry with dimensions is shown in Figure 1.

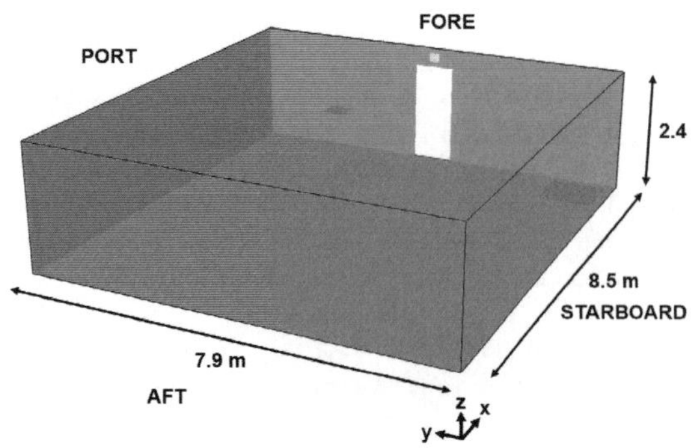

Figure 1: Room geometry with dimensions.

A door was located on the forward transverse bulkhead. The width and height of the door were 0.8m and 2.0m, respectively. The status of the door being open was varied in the simulations. An air exhaust was located above the door and a second air exhaust was on the same location on the opposite bulkhead. One of the air exhausts is shown in green in Figure 1, while the second one is not shown. An air supply was located in the middle of the room slightly below the ceiling level, shown in blue in Figure 1. The red square in Figure 1 shows the position of parametric t^2 design fires.

2.2 SURFACES

Three different bulkheads were studied in the simulations: A-60 class steel bulkheads, FRP bulkheads protected with mineral wool, and FRP bulkheads protected with an intumescent coating. The decks below and above the space, i.e., the floor and the ceiling, were defined as A-60 class steel sandwiches in all simulations.

All A-60 class steel structures were defined as sandwich structures consisting of three layers. The external layers were steel and the core was mineral wool. The FRP bulkheads were similarly defined to be sandwich structures with three layers. The external layers were made of FRP material under the trade name SAERTEX LEO®. The core of the sandwich was mineral wool. The protective layer was either an additional external layer of mineral wool or an external layer of intumescent coating. The protective layers were located on the bulkhead side that is inside the studied enclosure.

Three different combinations of bulkhead structures were studied in the simulations. The combination of the bulkheads was one of the following:

- all A-60 class steel, "steel" bulkhead combination,
- all FRP with intumescent coating, "coated FRP" bulkhead combination,
- three of the bulkheads FRP insulated with mineral wool and the fourth bulkhead FRP with intumescent coating, "partially insulated FRP" bulkhead combination.

The bulkhead structures of the different bulkhead combinations and their positions are summarized in Table 1. See Figure 1 for x and y coordinates.

Table 1: Bulkhead structures in the different bulkhead combinations. The structural layers are listed starting from the inside of the studied enclosure.

Bulkhead combination	**Structure**	**Bulkhead**
Steel	2 mm steel + 6 cm mineral wool + 2 mm steel	All
Coated FRP	2 mm coating + 2.9 mm FRP + 4.4 cm mineral wool + 2.9 mm FRP	All
Partially insulated FRP	2 mm coating + 2.9 mm FRP + 4.4 cm mineral wool + 2.9 mm FRP	y = 0.0 m
	6 cm mineral wool + 2.9 mm FRP + 4.4 cm mineral wool + 2.9 mm FRP	y = 7.9 m x = 0.0 m x = 8.5 m

2.3 MATERIAL MODELS

The required material properties for the simulations include density, conductivity, specific heat, and emissivity. As the fibre-reinforced polymer was assumed to thermally decompose in the simulations due to the elevated temperatures, a pyrolysis model was needed in addition to the material properties.

2.3 (a) Steel

The material properties of steel corresponded to the properties given for stainless steel in the Eurocode 3, which is the harmonised European standard for design of steel structures [2].

2.3 (b) Mineral wool

The material properties of mineral wool corresponded to general-type stone wool. Material properties of such materials are presented for example in [3].

2.3 (c) Fibre-reinforced polymer

The FRP material properties and the pyrolysis model were based on the experimental results acquired during the FIBRESHIP project. The development of the material model followed the modelling principles presented in [4].

The fibre-reinforced polymer was modelled as consisting of vinyl ester resin, glass fibre, and moisture. The mass of FRP material was assumed to consist of 23.75% of vinyl ester resin, 1.25% of moisture and 75% of glass fibre.

The vinyl ester resin was assumed to consist of two components based on the small-scale thermogravimetric analysis (TGA). The mass of the vinyl ester resin was assumed to consist of 42% of the first resin component and 58% of the second resin component.

The material properties of the vinyl ester resin components were assumed to be the same. The material properties of the FRP components are presented in Table 2.

Table 2: Material properties of the components of fibre-reinforced polymer.

Resin components 1 and 2	
Emissivity	0.9
Density	1000 kg/m^3
Specific heat	2.152 kJ/kgK
Conductivity	0.25 W/mK
Glass fibre	
Emissivity	0.9
Density	2400 kg/m^3
Specific heat	1.2 kJ/kgK
Conductivity	0.65 W/mK
Moisture	
Emissivity	0.9
Density	1000 kg/m^3
Specific heat	4.0 kJ/kgK
Conductivity	1.2 W/mK

The resin components were assumed to pyrolyse in elevated temperatures. The complex pyrolysis model of FDS was utilized to model the thermal degradation of the material [1]. A simplified presentation of the assumed pyrolysis reaction mechanisms is shown in Table 3. The reaction rates are dependent on the temperature, and some of the reactions are oxidative, i.e., the reaction rates are dependent on the local oxygen concentration. The produced fuel gas was assumed to be propane, which has a heat of combustion of approximately 44.6 MJ/kg. The produced inert gas was assumed to be water vapour.

The cone calorimeter results, for both a vinyl ester resin specimen (cured resin) and a specimen consisting of vinyl ester resin and glass fibre (laminate), were utilized to manually estimate the material properties for the glass fibre, the assumed resin components and their solid pyrolysis products. The experimentally measured density was used as a boundary value for the estimated component densities. The material properties were evaluated using expert judgement and similar reference materials to ensure that realistic values were used in the simulations.

Table 3: Reaction mechanisms of resin components.

Component	**Reaction no.**	**Products (yield %)**
Component 1	1	Solid product 1 (10 %) Fuel gas (90 %)
	2	Solid product 1 (80 %) Fuel gas (20 %)
Component 2	1	Solid product 1 (20 %) Fuel gas (80 %)
Solid product 1	1	Solid product 2 (8 %) Fuel gas (92 %)
Solid product 2		

Regarding the assumptions made about the other components of the FRP material, moisture changes phase into water vapour in elevated temperatures. The glass fibre was not considered reactive.

2.3 (d) Intumescent coating

The material properties and the pyrolysis model for the intumescent coating were based on the experimental results acquired during the FIBRESHIP project [5]. The development of the material model follows the modelling principles presented in [4].

Based on the TGA results, the intumescent coating was assumed to consist of two components. The TGA results have been reported in [5]. The material properties of the intumescent coating components were assumed to be the same and are presented in Table 4.

Table 4: Material properties of intumescent coating components.

Intumescent coating components	
Emissivity	1.0
Density	1500 kg/m^3
Specific heat	1.0 kJ/kgK
Conductivity	0.6 W/mK

The intumescent coating components were assumed to pyrolyse in elevated temperatures. The complex pyrolysis model of FDS was utilized to model the thermal degradation of the material [1]. A simplified presentation of the assumed pyrolysis reaction mechanisms is shown in Table 5. The produced fuel gas was assumed to be propane, which has heat of combustion of approximately 44.6 MJ/kg. The produced inert gas was assumed to be water vapour.

Table 5: Reaction mechanisms of intumescent coating components.

Component	**Reaction no.**	**Products (yield %)**
Component 1	1	Inert gas (100 %)
Component 2	1	Solid product 1 (94.6 %) Inert gas (5.4 %)
Solid product 1	1	Solid product 2 (40 %) Fuel gas (60 %)

	2	Solid product 3 (90 %) Fuel gas (10 %)
Solid product 2	1	Solid product 4 (50 %) Fuel gas (50 %)
	2	Solid product 5 (85 %) Fuel gas (15 %)
Solid product 3	1	Solid product 5 (10 %) Fuel gas (90 %)
	2	Solid product 4 (50 %) Fuel gas (50 %)
Solid product 4		
Solid product 5	1	Inert gas (50 %) Solid product 4 (50 %)

2.4 FIRE SCENARIOS AND DESIGN FIRES

Twenty-four (24) different fire scenarios were simulated. The design fire, the bulkhead combination, and the ventilation were varied to produce different fire scenarios as follows:

- four (4) design fires: parametric t^2 fire with slow growth rate, parametric t^2 fire with fast growth rate, a refrigerator fire, and an office furniture fire,
- three (3) bulkhead combinations: steel, coated FRP, and partially insulated FRP (see Table 1),
- two (2) natural ventilation conditions: leakage and open door (see Section 2.5).

The design fires in the simulations were defined as prescribed heat release rates as a function of time. The simulated physical time is 60 minutes for all fire scenarios.

In the simulations with FRP bulkheads, the FRP material pyrolysed, i.e., generated fuel gas, after reaching a sufficiently high temperature. The produced fuel gas ignited if there was enough oxygen available.

In the simulations, the fire began with the ignition of "a burner" with a defined area, describing a parametric t^2 fire, a refrigerator fire, or an office furniture fire. The heat release rate of the burner developed according to a defined curve of heat release rate as a function of time. In the beginning, the surfaces of the fire compartment did not contribute to fire. The combustible surfaces, i.e. FRP with intumescent coating, were assumed to pyrolyse and generate fuel in elevated temperatures. Their thermal degradation was modelled utilizing the pyrolysis model of FDS, with material properties presented above.

In this paper, the performance of FRP structures and protective material layers are discussed on the basis of the results obtained in the office furniture fire simulations. Therefore, the parametric t^2 fires and the refrigerator fire are described below only briefly whilst the office furniture fire is described in more detail.

2.4 (a) Slow and fast growth rate t^2 fires

The heat release rate (HRR) of t-squared (t^2) fires was described as $HRR = \dot{Q}_0(\frac{t}{t_0})^2$, where $\dot{Q}_0$=1000 kW, t is time in seconds and t_0 is 600 s and 150 s for the slow and fast t^2 fire, respectively. As the fire shortly became under-ventilated, it was decided to limit the HRR of the design fire by setting a maximum threshold. The maximum HRRs for the simulations of the rooms with closed door and open door were 5000 kW and 2000 kW, respectively.

2.4 (b) Refrigerator fire

The burning refrigerator was assumed to be 1.6 m high, 0.7 m wide and 0.6 m deep. The room was otherwise empty. According to Hietaniemi et al. [6], the maximum heat release rate of a refrigerator fire is roughly 2100 kW. In the simulation, the worst-case scenario was considered and thus the refrigerator fire was defined to have a maximum heat release rate of 4200 kW. The fire grew according to a slow t^2 fire growth rate until it reached its maximum heat release rate. After this, the heat release rate started to descend.

2.4 (c) Office furniture fire

The office furniture fire consisted of an explicitly defined workstation fire and modelled office furniture which took part in the fire. The explicitly defined workstation fire followed closely the experimental heat release rate reported by Kakegawa et al. in [7]. The modelled office furniture was assumed to have a similar heat release rate per unit area as a 20mm thick pine board in cone calorimeter tests of [8] under 50 kW/m^2 irradiance.

The furniture arrangement in the room and the location of the burning piece of table with explicitly defined HRR are presented in Figure 2. Figure 2 shows the fire development 320 seconds after the ignition. The heat release rate of the workstation is shown in Figure 3.

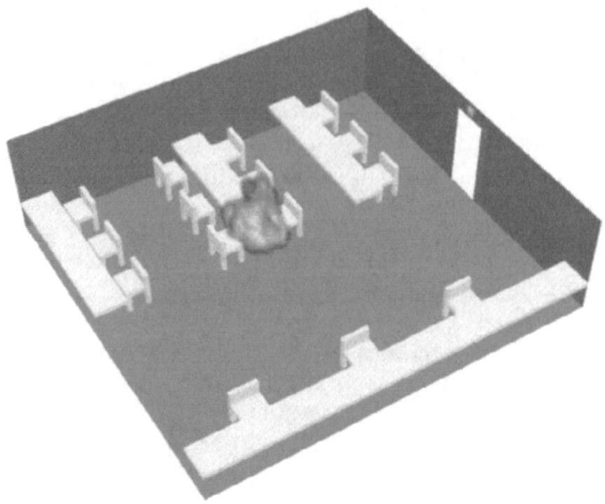

Figure 2: Office furniture fire scenario at t = 320 s.

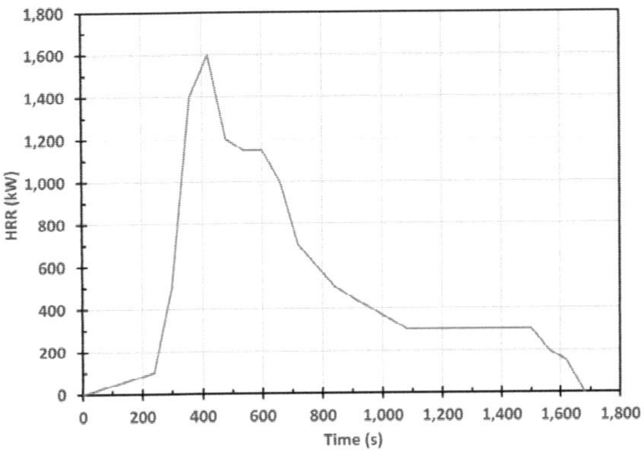

Figure 3: Heat release rate of a workstation.

2.5 VENTILATION

The accommodation space was assumed to be constantly mechanically ventilated in the simulations. The utilized ventilation system model did not take into account the pressure changes in the room caused by the fire. A ventilation rate of 1000 m^3/h was used in the simulations. The ventilation rate was estimated based on the design occupancy of the space and the minimum airflow rate per person as given by ISO 7547 [9]. As per the ISO 7547 design values for a saloon, the used design occupancy is 35 persons [9].

The natural ventilation of the space was modified in the simulations by opening the door in the forward transverse bulkhead. Each design fire was simulated with the door being either open or closed during the whole simulation. The open door was assumed to be connected to a space with constant ambient conditions. This is a conservative assumption, while fresh air was drawn to the space during the fire as the hot gases in the fire room expanded and exited through the door. This increased the heat release rate, because lack of oxygen did not limit the fire.

The enclosure boundaries were assumed to have some leakage, e.g., due to the penetrations through the boundaries. The total leakage area in the simulations with closed door was 0.05 m^2. In the simulations with open door, this leakage boundary condition was not included in the model. When the pressure in the space was below the ambient pressure, 101 325 Pa, fresh air was drawn to the space through the leaking area.

2.6 INSTRUMENTATION

The following quantities were monitored in the simulations:

- gas temperatures and oxygen volume fractions at discrete locations around the room,
- solid temperatures at discrete locations around the room on the bulkheads and the decks,
- mass and volume flows at all air supply and exhaust vents, and mass and volume flows through the door, if it was present in the simulation,
- temperature, visibility, oxygen volume fraction and velocity fields at both transverse and longitudinal planes going through the centre of the room,
- adiabatic surface temperature, burning rate, solid temperatures, and surface density fields at the boundaries of the domain.

3. SIMULATION RESULTS AND ANALYSIS

The results of the simulations with office furniture design fire are presented below as simulated average gas temperatures, simulated temperatures on the front of the first composite layer, potential heat release values, and fuel generation times of bulkheads. Potential heat release is the amount of energy (kWh) released if enough oxygen would be available for complete combustion of produced fuel gas. In the calculation, the produced fuel gas (kg) was scaled by the heat of combustion of the fuel (kJ/kg). The fuel generation times of various bulkheads, i.e. the start and end times of fuel generation, were determined on the basis of the mass loss rate visualisations of the simulations (see Figures 6 and 7).

This section aims to compare the simulations with the same design fire, and highlights the differences caused by the different bulkhead materials and structures.

3.1 SIMULATIONS WITH A CLOSED DOOR

The simulated average gas temperatures during an office furniture fire with a closed door are presented in Figure 4. The coated FRP bulkhead combination gives the lowest average gas temperature, indicating that the insulation capability of the FRP bulkheads with intumescent coating is inferior to both steel bulkheads and the FRP bulkheads with additional mineral wool layer. In the simulation with the coated FRP bulkhead combination, the furniture re-ignition that was present in the other two simulations did not occur. This is assumed to be due to the reduced gas temperature, caused by the greater heat loss through the bulkheads.

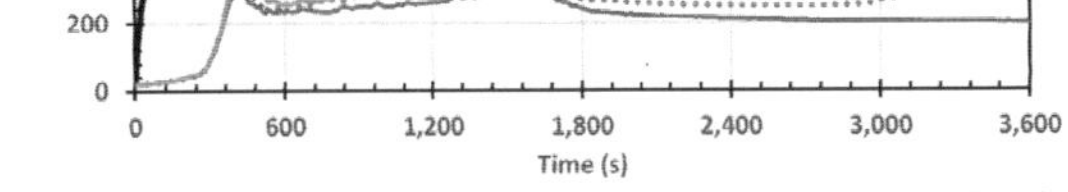

Figure 4: Average gas temperatures with office furniture design fire, door closed.

In the simulations of the office furniture fire with a closed door, the potential heat releases were approximately

530 kWh with the steel bulkhead combination, 380 kWh with the coated FRP bulkhead combination and 2600 kWh with the partially insulated FRP bulkhead combination. The potential heat release in the simulation with the steel bulkhead combination was almost 80% smaller than in the simulation with the partially insulated FRP bulkhead combination but approximately 40% greater than in the simulation with the coated FRP bulkhead combination. There are two causes for these differences: the different insulation properties of the bulkheads, and the additional fire load due to the combustible FRP bulkheads.

The simulated temperatures on the front of the first composite layer are presented in Figure 5. The presented monitoring points were located on two different bulkheads. In the simulation with the partially insulated FRP bulkhead combination, the monitoring point referred to as "fy0" was located on a bulkhead with protective layer of intumescent coating and the monitoring point referred to as "fxm" was located on a bulkhead with additional mineral wool layer. The effect of the improved thermal protection can be clearly observed in Figure 5, as the temperature on the bulkhead with intumescent coating is over 500% higher than in the bulkhead with additional mineral wool. On the simulation with the coated FRP bulkhead combination, the temperatures on both bulkheads were very similar during the whole simulation.

The fuel generation of the composite bulkheads was evaluated on the basis of the burning rate of the boundaries of the simulation domain, as illustrated in Figures 6 and 7. The fuel generation times of each composite bulkhead are presented in Table 6. The extent of pyrolysing area at the time of the first bulkhead starting to generate fuel is shown in Figure 6 and at the time of the last remaining bulkhead starting to generate fuel is shown in Figure 6. The bulkheads generated fuel until the end of the simulation. The bulkhead nearest to the workstation fire began to pyrolyse first.

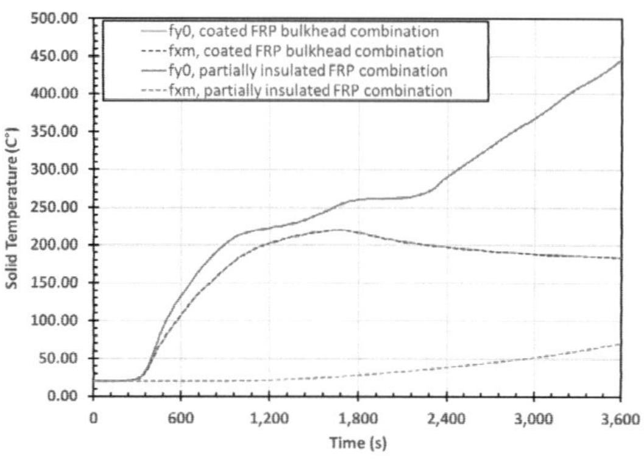

Figure 5: Simulated temperatures on the front of the first composite layer in the office furniture fire, with a closed door.

Table 6: Fuel generation times of the FRP room bulkheads, with a closed door, in the office furniture fire.

Bulkhead	Coated FRP room		Partially insulated FRP room	
	Start (s)	End (s)	Start (s)	End (s)
x = 0.0 m	780	–	–	–
x = 8.5 m	840	–	–	–
y = 0.0 m	900	–	780	–
y = 7.9 m	600	–	–	–

Despite the limited accuracy of the simulations in regards to the structural response, the simulated temperatures on the front of the first composite layers can be used as indicators for the integrity of the bulkhead. As can be observed in Figure 5, the temperature has increased above 100 °C in both monitored bulkheads in the simulation with the coated FRP bulkhead combination and in the bulkhead with intumescent coating in the simulation with the partially insulated FRP bulkhead combination. Only the bulkhead with an additional mineral wool layer in the simulation with the partially insulated FRP bulkheads has not reached the approximate glass transition temperature by the end of the simulation.

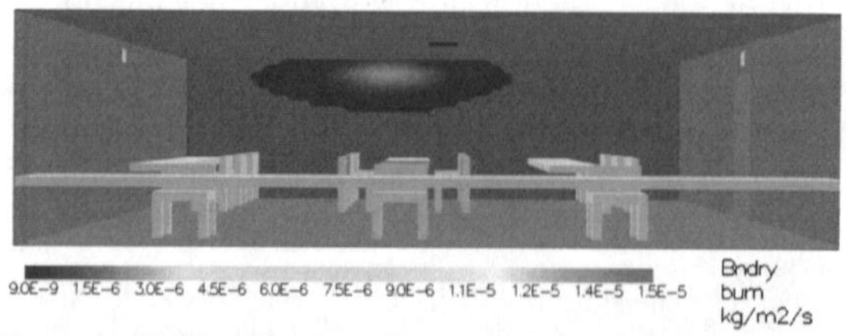

Figure 6: Fuel generation of composite bulkheads of coated FRP room, door closed, with office furniture fire, at t = 600s.

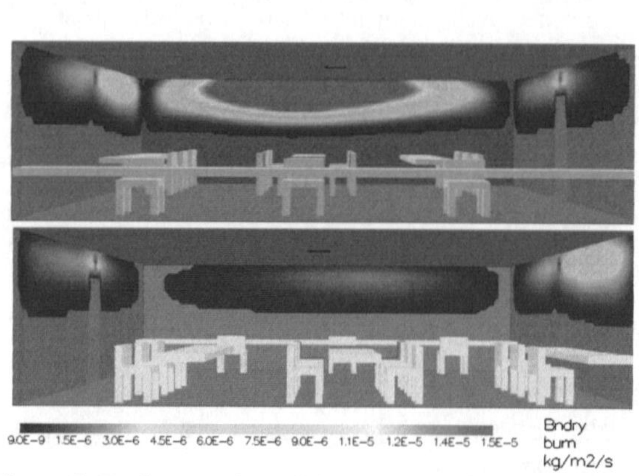

Figure 7: Fuel generation of the composite bulkheads of the coated FRP room, with a closed door, in the office furniture fire, at t = 900s.

At the end of the simulation, the bulkhead temperatures exceed the approximate glass transition temperature by 80 to 350%, which suggests that the integrity of the bulkheads could be compromised. Further studies would be required to estimate the structural deformation and its effect to the fire development.

3.2 SIMULATIONS WITH AN OPEN DOOR

The simulated average gas temperatures are presented in Figure 8. In the simulations with an open door, the gas temperature increased fastest in the simulation with the partially insulated FRP bulkhead combination due to the good insulation properties of the FRP bulkheads with additional mineral wool. The steel bulkheads also had better insulation properties than the coated FRP bulkheads, causing the gas temperature to be higher. Similar behaviour was observed in the simulations with a closed door.

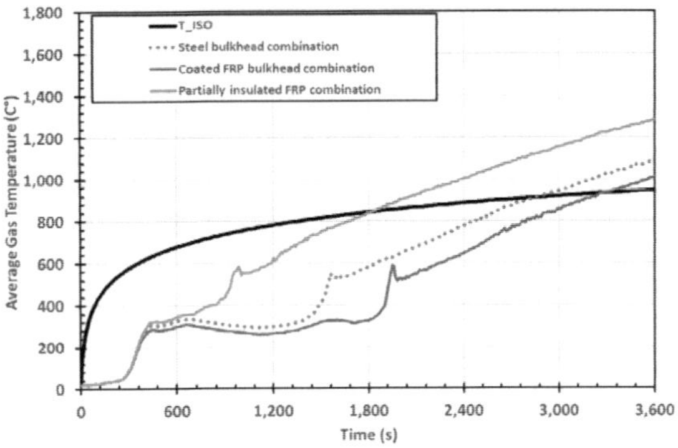

Figure 8: Average gas temperatures with office furniture design fire, door open.

In the simulations of the office furniture fire with an open door, the potential heat release rates were approximately 3200 kWh with the steel bulkhead combination, 5500 kWh with the coated FRP bulkhead combination and 4700 kWh with the partially insulated FRP bulkhead combination. The difference between the steel and the coated FRP bulkhead combinations was approximately 70% and the difference between the steel and the partially insulated FRP combinations was approximately 50%.

The difference of 15% between the potential heat releases in the simulations with the coated and the partially insulated FRP bulkhead combinations can be explained with the fuel generation times of the bulkheads. The fuel generation times are presented in Table 7. In the simulation with the coated FRP bulkhead combination, all bulkheads began to pyrolyse within the first 720 seconds (12 minutes), as the intumescent coating did not provide sufficient protection from the elevated temperatures in the space. In the simulation where the bulkheads had a protective layer of mineral wool, the last bulkhead to pyrolyse did not begin to generate fuel until at 3360 seconds (56 minutes). In the simulation with the steel bulkhead combination, there was no additional fire load provided by combustible bulkhead material, which caused the potential heat release to be smaller.

Table 7: Fuel generation times of FRP room bulkheads, door open, with office furniture fire.

Bulkhead	Coated FRP room		Partially insulated FRP room	
	Start (s)	End (s)	Start (s)	End (s)
x = 0.0 m	600	–	3360	–
x = 8.5 m	720	–	2880	–
y = 0.0 m	660	–	600	–
y = 7.9 m	480	–	2880	–

The simulated temperatures on the front of the first composite layer are presented in Figure 9. The presented monitoring points were located on two different bulkheads. In the simulation with partially insulated FRP bulkhead combination, the monitoring point referred to as "fy0" was located on a bulkhead with a protective layer of intumescent coating and the monitoring point referred to as "fxm" was located on a bulkhead with an additional mineral wool layer. The effect of the improved thermal protection can be clearly observed in Figure 9, as the temperature on the bulkhead with an intumescent coating is over 200% higher than on the bulkhead with additional mineral wool. In the simulation with the coated FRP bulkhead combination the temperatures on both bulkheads were very similar during the whole simulation.

In the simulations with a closed door, the simulated temperatures on the front of the first composite layers can be used as indicators for the integrity of the bulkhead. As can be observed in Figure 9, by the end of the simulation the temperature has increased above 100 °C in all the monitored bulkheads.

At the end of the simulation the bulkhead temperatures exceed the glass transition temperature by approximately 200 to 800%, which suggests that the integrity of the bulkheads could be compromised. Further studies would be required to estimate the structural deformation and its effects to the fire development.

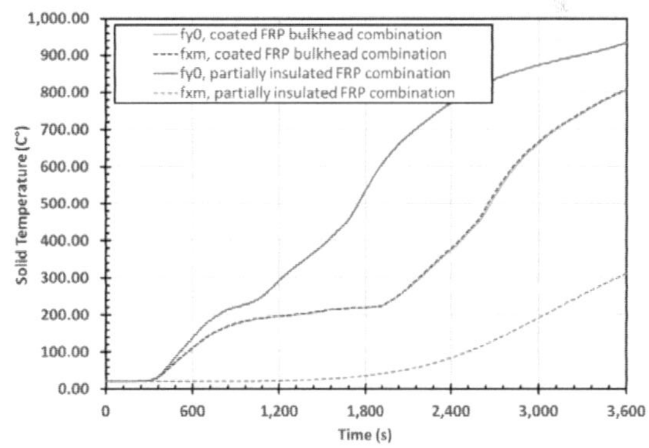

Figure 9: Simulated temperatures on the front of the first composite layer in the office furniture fire, with an open door.

3.3 SUMMARY

On the basis of the simulation results, it is concluded that the additional layer of mineral wool is more effective in thermally protecting the bulkheads than the intumescent coating. Improved insulation protects adjacent spaces from the heat and thus increases the gas temperature in the space in question.

In addition, the additional layer of mineral wool will either prevent or postpone the pyrolysis of the bulkhead, depending on the fire exposure. In the simulations with an open door, the potential heat release rate in the simulations with the FRP bulkheads were 50 and 70% greater than in the simulation with the steel bulkheads. The potential heat release rate was the largest for the simulation with the protective layer of intumescent coating.

In the simulations with a closed door, the potential heat release rate with the partially insulated FRP bulkheads was approximately 400% greater than in the simulations with the steel bulkheads. This is due to the fuel gas which was produced by the bulkheads. However, in the simulations with a closed door and the coated FRP bulkhead combination, the potential heat release was reduced by approximately 30% in comparison to the simulation with the steel bulkheads. This is due to the increased heat loss through the bulkheads, which alters the fire behaviour.

The temperature increase above the glass transition temperature was used as an indicator for significant loss of load carrying capacity. Significant loss of load carrying capacity could mean that the integrity of the bulkheads is compromised. When the door was closed in the simulations, the load carrying capacity of all bulkheads without mineral wool protection could be compromised. When the door was open in the simulations, the load carrying capacity could be compromised also for the bulkhead with mineral wool protection.

4. CONCLUSIONS

Based on the simulation results, FRP materials need to be well insulated to restrain the temperature increase in fire conditions, and the amount of insulation required for compliance with IMO FTP Code Part 11 test should be further assessed. If the material temperature is allowed to increase, the material will begin to contribute to fire and the structure can potentially lose its integrity. In the simulations, the bulkheads that were protected only with the studied 2mm intumescent coating exceeded the material's glass transition temperature after heat exposure equivalent to only 11 minutes of the standard ISO 834 time-temperature exposure. When the bulkhead was protected by a 60mm layer of mineral wool instead, the glass transition temperature was exceeded after heat exposure equivalent to 16 minutes of the standard ISO 834 time-temperature exposure.

The results presented in this paper are valid only for the specific protective solutions studied with the layer thicknesses used in the simulations. Further studies are needed for exploring the effect of layer thickness variations on the protective capability.

5. ACKNOWLEDGEMENTS

This project has received funding European Union's Horizon 2020 research and innovation programme under grant agreement No. 723360.

6. REFERENCES

1. K. McGrattan, S. Hostikka, R. McDermott, J. Floyd, C. Weinschenk, and K. Overholt, "Fire Dynamics Simulator Technical Reference Guide Volume 1: Mathematical Model - NIST Special Publication 1018 Sixth Edition." National Institute of Standards and Technology, Gaithersburg, Maryland, p. 173, 2013.

2. European Committee for Standardization, EN 1993-1-2:2005 Eurocode 3: Design of steel structures. Part 1-2: General rules. Structural fire design. 2005.

3. M. J. Hurley (ed.), "SFPE Handbook of Fire Protection Engineering, Fifth Edition." Springer, New York, NY, 2016.

4. A. Matala, "Methods and applications of pyrolysis modelling for polymeric materials," VTT Science 44. VTT Technical Research Centre of Finland, 2013.

5. FIBRESHIP Deliverable D2.4 (WP2), "Report and database on the results of the fire performance experiments." 27/09/2018. 46 p. + app. 26 p. To be published in 2020.

6. J. Hietaniemi, J. Mangs, and T. Hakkarainen, "Burning of electrical household appliances: An experimental study," VTT Research Notes 2084. VTT Technical Research Centre of Finland, 2001.

7. S. Kakegawa, Y. Yashiro, H. Satoh, H. Kurioka, I. Kasahara, Y. Ikehata, N. Saito, and T. Turuda, "Design fires for means of egress in office buildings based on full-scale fire experiments," Fire Safety Science – Proceedings of the seventh international symposium, pp. 975–986, 2003.

8. ISO 5660-1:2015. Reaction-to-fire tests – Heat release, smoke production and mass loss rate – Part 1: Heat release rate (cone calorimeter method) and smoke production rate (dynamic measurement). Third edition 2015-03-15. Geneva: International Organization for Standardization, 2015. 55 p.

9. International Organization for Standardization, ISO 7547:2002 Ships and marine technology. Air-conditioning and ventilation of accommodation spaces. Design conditions and basis of calculations. 2002.

7. AUTHORS' BIOGRAPHY

Alexandra Tissari holds the current position of Research Scientist at VTT. Her research activities focus on fire simulations. She has a background in naval architecture.

Timo Korhonen holds the current position of Senior Scientist at VTT. He is experienced in the field of fire research including fire simulations, development of fire risk analysis methods, development of evacuation models, and expert consultation tasks in the area of fire safety. In addition to the research work, he has lectured on fire safety courses at Helsinki University of Technology (academic years 2007-2008, 2009-2010, 2011-2012).

Terhi Kling holds the current position of Senior Scientist at VTT. She has lengthy experience in fire risk assessment using numerical simulation of smoke transport and human evacuation. She has used the FDS code extensively on ship, train and land applications and carried out research and development in several international projects.

Antti Korkealaakso holds the current position of Research Trainee at VTT and is a third-year student in mathematics and systems sciences at Aalto University. He is responsible for fire simulations.

Tuula Hakkarainen holds the current position of Principal Scientist at VTT. She is experienced in the assessment of fire performance of materials and products for buildings and vehicles. In addition to the research work, she has acted as the project manager in several international and domestic projects related to fire safety of surface transport, e.g. SURSHIP-FIRE, TRANSFEU and FIBRESHIP.

A 10-DAY COURSE TO PLAN AND EXECUTE FIELD STUDIES FOR MARITIME DESIGN PROCESSES

E. Gernez and **K. Nordby**, The Oslo School of Architecture and Design, Norway

SUMMARY

Human-centred design processes often include field studies to achieve effective design solutions that support user needs. Despite this, field studies are rarely used in maritime design because they are time consuming and because it is hard to ensure that the field study outcome will have an impact on the design process. One way of overcoming these challenges is to better train designers in how to carry out efficient field studies aligned to the needs of ongoing design processes. We present a 10-day course to plan and execute field studies. The students' feedback indicates that the students are able to identify key challenges in their field study practice and propose solutions to address them. We conclude with propositions for how to improve the course and suggest to open it to a wide audience of designers and engineers with little or no experience with field studies.

1. INTRODUCTION

Field studies come from the ethnographic tradition of the human and social sciences (Blomberg et al., 2009). In such studies, field researchers live in the field together with field informants, often for an extended period of time. This enables the researchers to gain a first-hand personal, physical and transformational experience from which they can derive insights. Originally performed by ethnographers and anthropologists, field studies are now a regular part of human-centred design processes. Such processes, however, do not directly represent anthropology and are usually done in a much shorter time span and focused on specific design needs (Kujala et al., 2003). Because of this, field studies in design have been labelled 'ethnomethodology' (Button, 2000) or 'problem-oriented ethnography' (Lützhöft, 2004).

In a design process, field studies can create a deep understanding of a given group of users, their context and their tasks, so that products and services can be designed that will be more likely to be well understood, adopted and used to satisfy their needs (Kujala, 2003). There are, however, no guarantees that this will happen, and the specific impact of a field study on a design process cannot be known in advance. Field studies are time demanding, partly because they often generate a vast amount of (mostly qualitative) data that requires time to process. The ability to target specific data to be collected, to select a collection method that will capture this data with both high fidelity and time efficiency, and document the data efficiently are all important success factors of a field study.

The data collected in the field needs to undergo a transformation before becoming usable data for the design process. First, the insights collected are the result of the individual, physical experience of one designer. Individual insight becomes team insight after each designer has shared his or her experiences with the members of the design team, including those who have not been in the field. Team insights become data for the design process when it is connected to a specific user need in a specific context of use.

In addition, it is a challenge to organise a field study. Gaining access to the field and its informants requires time and good connections in order to build a relationship of trust and openness. Some contexts might be safety critical and require the field researcher to go through safety courses beforehand. Because of weather uncertainties, field studies on board ships can be delayed, and the field researcher might need to spend more time than originally planned on board the ship (Lurås et al., 2015).

We can summarise the challenges of using field study methodology as understanding how to gain access to the field and its informants, knowing how to obtain the right data, and doing so efficiently. Field data needs then to evolve from a personal and physical experience to team insights derived from collaborative reflection and analysis. In order to overcome these challenges, it is important that designers are properly trained and experience the entire process of planning and doing field studies for design processes for themselves.

2. THE COURSE

The goal of the 10-day field study course is to present the students with a method to deal with all the challenges of field study methodology. The course was developed as a part of the ONSITE research project in which we are designing and carrying out field studies with Norwegian maritime industrial partners. Partners include a ship design and building company, a company designing and assembling engine rooms for ships, and a safety and efficiency consultancy company. With the course, we transfer our experience with ONSITE and previous field studies in the maritime industry to the students (Lurås and Nordby, 2014, 2015).

Field studies are useful in the design process when they are time and resource efficient and built to deliver targeted opportunities for innovation that can be used in ongoing or new design processes. In order to deliver this quality, we have derived a set of guidelines for the field study planning process:

- *Human centeredness*: The field study aims to uncover user needs in a way that fully and ethically respects the privacy and integrity of the users.
- *Desired impact*: The process begins by agreeing with the customers about how the field study will contribute to competence building, innovation, and the design of a specific product or service.
- *Clear focus*: An agreement must be reached with the customers as to where the field study will take place and what will be observed there.
- *An implementation plan with room for improvisation*: The implementation plan contains instructions and directions enabling the researcher to improvise with confidence.
- *A strategy to communicate the field study observations*: The value of the field study lies in how insights are transferred. A significant share of the field study resources need to be allocated to this transfer.

We base the course content on the Lurås and Nordby model (Figure 1) of the three design activities embodied in the field study methodology:

- *Design proposal and reflection*: The students are asked to propose design concepts and present them in their final report.
- *Data collection*: Our set of guidelines touches upon the importance of documentation efficiency, and we introduce methods for efficient observation and documentation.
- *Field experience and reflection*: The students are given assignments that emphasize the importance of reflection as a significant part of their field experience.

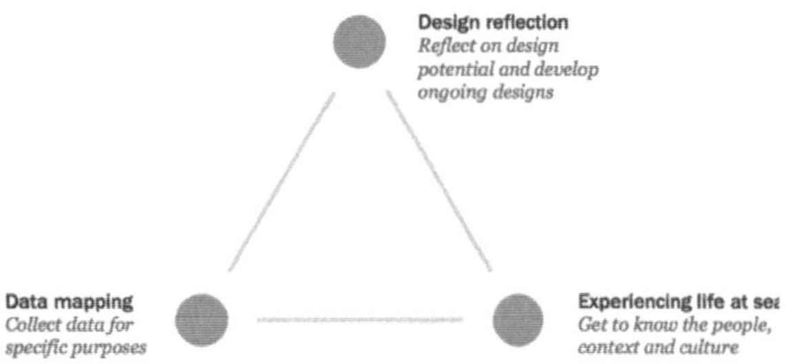

Figure 1: Lurås and Nordby´s design-driven field research model (reproduced with authorization from Lurås and Nordby, 2014).

Further, we base the course structure and learning activities on the experiential learning processes developed by Kolb (1984) modelled upon Schon´s (1983) and Dewey´s (1938) work on how experiences and reflections on experiences shape learning. The main idea is to arrange for learning activities that provide students with experiences and the opportunity to reflect about these experiences in order to help them assimilate the associated learning. Lurås and Nordby's model is designed to follow this type of learning process. For other examples, see the review by Tonkinwise (2005) on the use of reflective and learning processes in design.

We define the following learning objectives for the course:

- Understand the importance and ethical dimensions of field studies for design
- Learn to plan effective field studies
- Learn to use the most important observation methods
- Learn to reflect on design processes while in the field
- Learn to organise data from field studies
- Learn to analyse field data, individually and as a team
- Learn to share insights from field studies
- Learn to write field study reports

The course has been held so far for students in their third year of a master's program in design. For the 2016 and 2017 editions, we took a group of approximately 20 students on a passenger ferry ship. Ships with a regular route and schedule make the logistics of carrying out the course relatively easy and predictable. Passenger ships also provide many sites for observation, most of them not too technical, allowing students with no background in ship design and maritime engineering to collect insights. In addition, large ferry ships allow us to distribute many students across the ship without interfering too much with the ship's and crew's daily operations.

The course overview is visualized in the Figure 2 below. The content of each day is explained in the next sections, detailing the activities proposed to the students, what material they are given, what assignments they are given, and what material they are asked to produce.

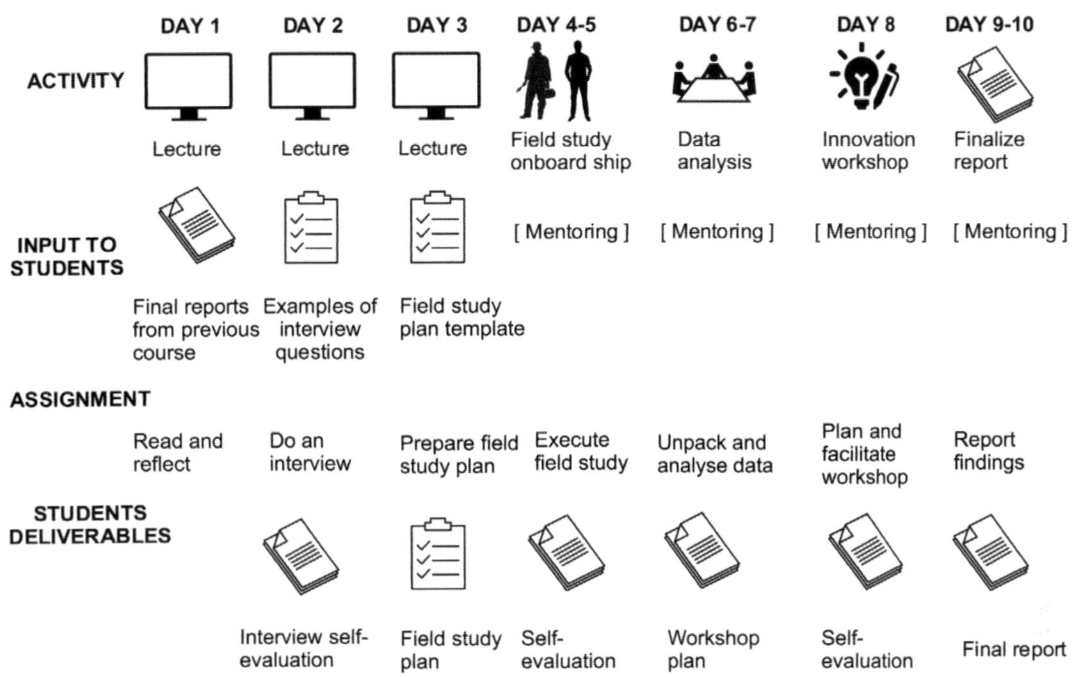

Figure 2: Overview of the day-by-day course activities.

2.1 DAY ONE – INTRODUCTION LECTURE

The first day began with a lecture where we shared our experience doing field studies on ships. The students were shown highlights from a design project that used field studies extensively (Lurås and Nordby, 2014). Using examples from this project, the students were introduced to the opportunities and challenges of sending designers into the field, sharing field insights among the design team, and collaboratively turning the insights into design objects. Examples of observations from the field were given along with ethical considerations regarding how these observations might be gained and used. We reminded the students that their work could have real economic consequences for the informants and for the companies involved.

The students were then given a selection of final reports produced by the students who took the course the year before. The objective was to give them an idea of what they would be able to produce at the end of the course and to make them aware of what type of final deliverable they would be asked to produce. The students were asked to read the reports and critique them in terms of what they thought was well done and communicated and what could have been improved upon.

2.2 DAY TWO – PROCESS AND METHODS

On the second day, we delved into the field study process and methods by way of a lecture. The field study process was covered in general terms and then in detail for each step. Observation methods such as ‘thinking aloud’, ‘artefact walkthrough’, ‘contextual inquiry’, and ‘context mapping’ were introduced. Finally, a specific lecture was given on interviewing techniques since mastering interviewing skills is a requirement for using most of the other field methods. In addition, interviews are very common in collaborative design and co-creation processes, so it is an important skill to practice.

The assignment for that day was to plan and perform a 30-minute audio-recorded interview with each other, which was then transcribed. Each student conducted and gave one interview. To ease the interview training, the students were asked to perform the interview in a familiar context where they do complex physical work. We used the school workshop as a location and context where the students were investigating how other students were using the facilities.

Importantly, after the interviews were done, the students were asked to listen to the audio again to evaluate themselves and ask: What went well? What did not? How could technique be improved? Both the transcript and the reflection are mandatory deliverables for the course.

2.3 DAY THREE – PRACTICAL ADVICE

The third day was the last day of lecture, and it focused on sharing very practical advice for how to go about field studies: how to get in touch with contacts that might provide access to the field, how to pitch the benefits of doing a field study to a potential customer, how to co-design and communicate the scope of the field study with a potential customer. Then, we reviewed the equipment to be taken into the field, how to prepare for note-taking, and how to rework the notes to capture insights. The students were given a number of digital templates that could be used in their study: a field study plan, a letter informing about the scope of the field study, an equipment checklist, a list of useful questions to ask during interviews on the field.

The assignment for this day was to prepare a field study plan and get ready for the upcoming field study. Arrangements had been made beforehand with a ship company to host the field study. The students were given templates of design briefs to help them define their desired focus for the field study, which could then be turned into a plan. We mentored the students regarding the elaboration of their plans; based on their interests, they were given tips on what to look at, and how to look at it.

2.4 DAYS FOUR AND FIVE – FIELD STUDY

The field study took place during these two days. This year, we boarded a passenger ship crossing from Norway and Sweden and returning the same day. The students spent approximately five hours on board. Together with the ship company, we identified beforehand areas of interests for the students: bridge, engine room, engine control room, kitchen, buffet restaurant, and duty-free shop. Two student groups looked at the 'customer experience' throughout the ship working. In most cases, it is best to use teams of two students so as to not overwhelm the users with too many field researchers observing and interacting with them. It is also best to make the students plan their individual roles during the field study: who will take notes, who will take pictures, who will ask questions.

The captain of the ship was aware of the field study and had given his green light beforehand. He knew where the students would be carrying out the field study, but he did not know exactly what they would be doing. The students were given the task of introducing their project to the ship's crew. They presented orally why they were there, what are they were interested in, what they needed from the crew. They also had the information in written form, ready to give to the crew if they requested it.

We were present during the field trip and stationed at one place on the ship, available to help and mentor the students. The students used the first half of the field study (Norway to Sweden) to perform a so-called explorative field study, where they familiarized themselves with the context and its users. On the second leg of the trip (Sweden to Norway), they focused on specific aspects and deepened their findings. After the field study, the students were asked to reflect on their experience and write a summary. This reflection is a mandatory deliverable.

In the past, our experience has been that students who choose to work from the bridge are the most likely to get sea sick because that is the highest point in the ship and where one is in constant eye contact with the sea. We therefore recommend to check the students' sensitivity to seasickness before allocating the focus areas. Some students this year got sick, but they still managed to record in audio most of what was happening around them while they were lying down, so they actually managed to collect very good data and work with it later on. Some students used seasickness tablets that made them a bit dizzy and reduced their ability to concentrate.

2.5 DAY SIX AND SEVEN – ANALYSIS

A short debriefing of the field study was done in plenum where students shared their experiences. Then the students were briefed and mentored in their analysis of their findings. During the rest of the day, we went from student group to student group to answer questions and give help. We observed that often the students had collected good material, but that it was challenging for them to assess the value of their findings and connect the findings to design problems. We tried to help them reflect over their findings and frame the design problems they wanted to work with.

The assignment for these two days was to prepare for a workshop where the findings would be communicated and worked on. The students were asked to deliver a plan for the workshop and prepare the materials for it including selecting and printing key visuals and creating new visuals specially designed for the workshop.

2.6 DAY EIGHT – WORKSHOP

Ideally, the workshop convenes the design team together with informants from the field and relevant stakeholders of the design process. The workshop is a very important step in a field study because it enables communication of individual field insights to the rest of the team, especially to the members who did not go on the field trip. It enables a collective review of the findings in order to validate and prioritize them, and it augments some of the most important findings with additional visualizations, ideas, and maybe even testing some ideas and sketching some prototypes.

For the course setup, it is a challenge to bring in field informants, and there is no specific design process connected to the field study. The workshops are held between two groups of students: one student group presents their findings the other group and invites them to collaborate in working with the findings; then the two groups swap roles. This way, the two most important 'post-field study' skills are being trained: transferring field knowledge inside a team and collaboratively working with the field knowledge. The students are asked to film their workshop for documentation purposes. We see both arranging the workshop and being a contributing part to an arranged workshop as important learning opportunities. The workshop format emphasizes the transfer of knowledge in field studies as not only recorded data, but as experiences that can be shared through conversation and collaboration. Again, the students were asked to write down and submit their reflections about their workshop experiences.

2.7 DAY NINE AND TEN – FINAL REPORT

For the last two days, the students prepared a report that assembled their findings and recommendations from the field. A suggested structure for the report is provided: (1) summary of findings (2) field study goals (3) field study methods (4) type of data collected (5) overview of the surveyed area with users and systems (6) observations and findings (7) framing of main findings with proposed innovation concepts.

We mentored the students throughout these two days. The final reports are submitted in Word format, and we usually share the reports with the ship owning company that granted access to their ship as a way to thank them.

3. COURSE EVALUATION

The objective of the course evaluation is to collect data about how the students experienced the course in order to improve structure and content. For example:

- Knowing what the students experienced as the most challenging aspect of the course helps us improve the course structure by prioritizing specific learning activities.
- Collecting input from the students on what they would propose as best practices for the field study methodology, based on their experience, helps us tailor the course content to this specific audience.

The course evaluation is based on collecting and analysing data about the experiences of the students. We collect two types of data:

- *Written reflections of students in text format:* The students are asked to reflect on specific learning activities and write down their reflections as part of course assignments. These reflections are submitted electronically through the school´s Course Management System (CMS), Moodle (Moodle, 2017)
- *Pictures and videos of the students performing the field study:* The students are asked to share all the pictures they took during their field study. They are also asked to video record and submit the workshops held after the field study.

The data set is built upon the three most fundamental activities for learning to plan and execute field studies: interview training on day three, field study on days four and five, and workshop on day seven. Extracts from the student reflections are translated from Norwegian into English. When needed, some quotes are slightly rephrased to be self-explanatory without needing additional context. The presented pictures are anonymised.

3.1 DATA ANALYSIS

We review the student´s written reflections against two perspectives:

- Challenges: What were the challenges identified by the students in this learning activity? This perspective helps us check to see if that what we think is challenging (and hence important to learn) matches the experience of the students. If yes, then it means that the learning activity is relevant and efficient and that the learning outcomes of the course are likely to be achieved.
- Best practices: Did the challenges identified by the students actually matter for delivering a high-quality field study process? This perspective helps us check whether or not the students have understood what makes the difference between good and bad practices. If yes, this means that the students are more likely to be on track with learning the fundamentals of field study methodology.

We review the picture-based data by looking for evidence of basic observation and documentation skills exhibited by the students.

3.2 EVALUATION RESULTS

3.2 (a) Interview training

The following types of challenges were observed by the students in this session of the course.

Getting the informant to talk:
'I asked questions that gave me only yes or no answers'.
'I could have talked less and commented less on the answers'.
'I could hear myself finishing the informant's sentences'.
'I found it challenging to give my informant enough time to think about the question, and give more time after to come up with additional reflections'.

Asking relevant questions:
'I felt my questions where not specific enough and that I had to rephrase them several times to get the type of answer I was looking for'.
'I could have made my questions better adapted to the topic of the interview'.
'I could have had several alternatives for follow-up questions'.

'I asked a very open question: "What could be different?" [...] I believe the answer I got would have been better had I had specified the relation to the parameter of my question'.
'I felt I was not good enough at using my informant's answers to progressively build more precise questions'.

Interview techniques, conversation facilitation material:
'I tried a "walk and talk" technique during the interview, and that worked well; that provided me with the most complete answers of the interview'.
'For the workshop, I sketched a map of where the interview took place and used it as a support to the interview. However, I could have planned better how to use the map in the interview'.

Planning and managing the interview:
'I followed my list of questions quite closely, and that deteriorated the flow of the conversation'.
'It did not feel natural to use ready-made questions from a template'.
'I liked the flow of the conversation, but I was unsure how far I could follow my informant's digressions with regards to my interview plan'.
'I tried to rephrase the answers I got to confirm that I understood correctly, and that worked well'.
'I tried to remember to tell my informant in the course of the interview that it was going well, that I was getting the information I needed, and to thank my informant for his/her time'.
'I could have had a more specific goal for the interview'.
'I could have focused my interview on a more specific problem'.

The examples show that listening to the interview recording made the students become immediately aware of important challenges that could help them improve their interview technique. For instance, how to interact with their informants, giving them time to think about and reflect on their answers, asking questions that progressively narrow down and deepen the topic, designing and using material to facilitate the conversation, and giving feedback to the informant. Based on our experience and on interview methodology literature (Kvale, 1996; Kvale and Brinkmann, 2009), all these identified challenges are very important success and quality factors.

3.2 (b) Field study

The following types of challenges were observed by the students.

Observation and documentation:
'I was very happy with our observations because we managed very early on to spot interesting problem areas, which helped us make decisions for what to observe among the vast information available around us'.
'It would have been more helpful to document a scene with video instead of photos. That would have shown the whole experience from A to Z, as well as how our informants behaved'.
'We did not ask if we could record the interviews. We felt our informants would be more sceptical about participating and that it might impact their answers. I feel we should have tried anyways because it would have helped us better collect and process the data'.

Interaction with the informants:
'Our informants needed further explanations about the goal of our field study and how they could help us do our field study'.
'Our informants seemed to be at ease with our presence and our work. We explained to them first why we were here, what we planned on doing and how we planned on doing it. It seems that helped them relax'.
'Our interviews looked more like conversations than structured interviews. We could have had more focus on what information we were trying to get from our informants'.

Overall planning and execution of the field study"
'The informants we talked to were regular customers who had not exactly experienced the problem we were interested in solving'.
'We stopped several times to go through our notes to update and refine the problem areas we had defined'.
'It's surprising how fast time goes'.
'What was most important for me to learn was not the field study itself but rather the importance of being prepared both before and during the field study. ... It's been very educational to see how good preparations can improve the findings and the experience of the field study'.
'The biggest challenge for me has been to define focus areas. However, it went quite well, thanks to a systematic review of our findings during the field study and mentoring from the course holders'.

The extracts above show that the students identified important success factors in line with our own experience, for instance, being very well prepared, being very clear with their informants, selecting the appropriate observation methods and the appropriate informants, and taking the time to review the findings while in the field.

Photos taken by students during field studies show some skills they were able to use during this learning activity (see Figures 3–6 below).

Figure 3: Interview of user in context.

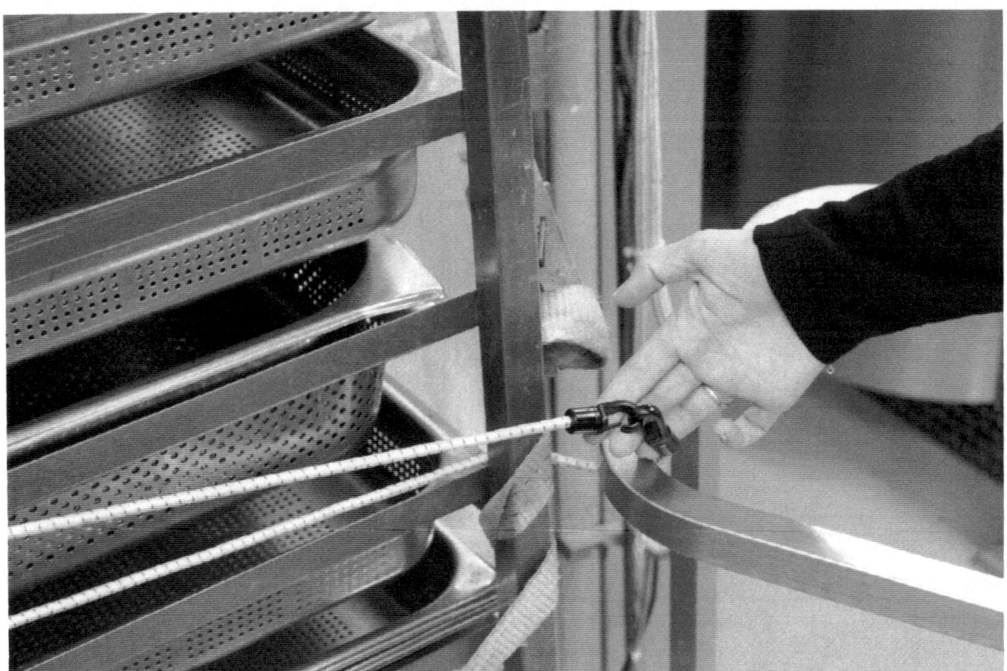

Figure 4: Identifying a potential issue (trip and fall involving cooking objects in the kitchen).

Figure 5: Reviewing notes and sketching while on the field.

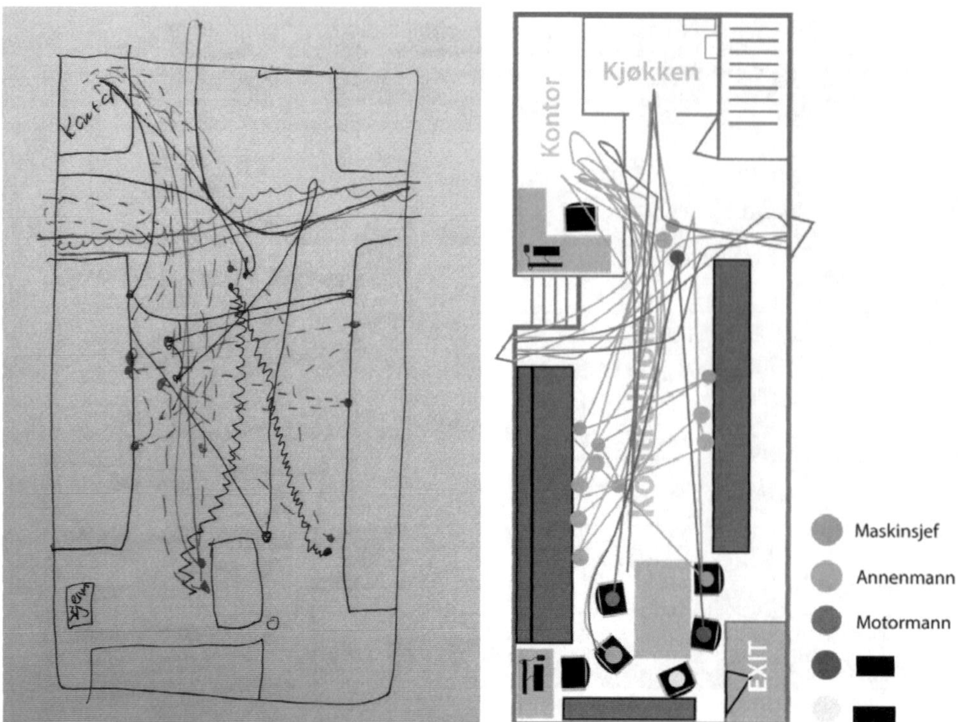

Figure 6: Sketching in the field (left) and output in field study report (right). The sketch is an attempt to document the movements of the users of an engine control room.

3.2 (c) Workshop

The following types of challenges were observed by the students.

Communicating field findings to team members who were not in the field:
'Putting forward to others what we found helped me see the big picture of our findings. It also helped us with deepening and consolidating the reasons we chose to focus on specific problem areas'.
'It was very useful and refreshing to talk to someone who could see our findings from a different perspective'.

Managing the engagement and participation in the workshop:
'Participants were initially interested in the problem areas we identified, but when we let the participants work with the problem areas by themselves, the motivation fell down'.
'We communicated very tightly defined problem areas. The workshop participants told us that they felt ideas were "created in their heads" before they could even think about it. That limited their ability to come up with new ideas'.
'A white page can seem intimidating because the participants might feel that they need to fill the whole page'.
'We were asked as participants to do several short and intense work tasks. That produced a lot of ideas, but a lot of the new solutions did not solve the problem because it was not clear what the workshop facilitators asked for'.
'It was very challenging to get the workshop participants to write down or draw instead of only talking'.

Preparing for the workshop:
'Preparing for the workshop forced us to discuss our findings with each other before presenting them to the rest of the team'.
'We should have done a dry-test of our workshop beforehand.... We interrupted each other often during our presentation of the findings'.
'I was not exactly sure what to ask the workshop participants to do because I was not sure what I wanted to get out of the workshop'.
'Next time I would like to be better at tapping into the competence of the workshop participants, get them to prepare themselves for the workshop, check what they think would be most useful for them'.

These extracts show that the students become immediately aware of important distinctions that can really improve the quality of their workshops, for instance, testing the workshop plan beforehand, designing a way to communicate results, designing activities to engage the team, and dealing with the ownership of ideas.

Screenshots extracted from the video documentation of the workshop showed some skills that the students have been able to use during this learning activity (see Figures 7–9 below).

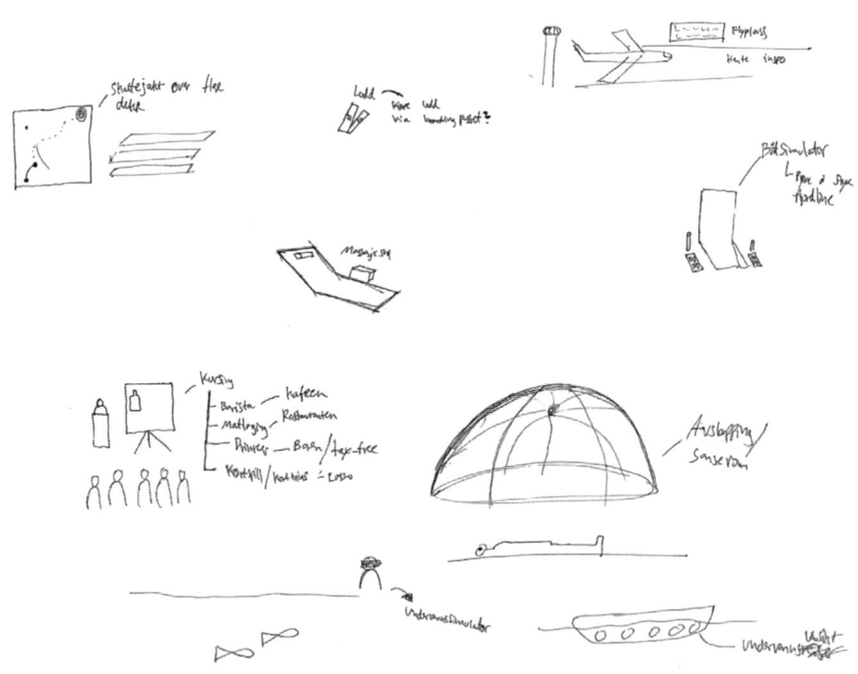

Figure 7: Concept ideation during workshop.

Figure 8: Results presentation and co-design in workshop. Pictures from the field study are presented on a screen, a reconstructed map of the studied area is put forward on the table, and the students use sketching while talking.

Figure 9: Results presentation and co-design in workshop. A map of the main observed issues is put forward on the table. Post-its and a white board are used to visualize and analyse short ideation tasks given to the participants.

4. DISCUSSION

Overall, the course seemed to be very useful for the students. One student wrote in the field study self-evaluation: 'It's been very educational to observe a context and users in their context, compared to what I have done previously, which was mostly interviews of users outside their context'.

This course had several limitations. First, it is especially useful to industrial design students who are already trained in human-centred design activities such as collecting insights from user behaviours. It would be interesting to test the course with engineering design students and maritime engineers working with the design of products and services in the maritime industry. They might have a perspective that is more technology-centred than human-centred. To explore this topic, we plan to build on research

of the introduction of human-centred design and human-factors concepts in technology-centred maritime design processes. Lutzhoft and Abeysiriwardhane as well as Salustri and Neuman look at the problem from an education perspective, while Lundh and Mallam look at it from an operational perspective (Abeysiriwardhane et al., 2015, 2017; Mallam et al., 2015; Mallam and Lundh, 2014; Salustri and Neumann, 2017).
Second, the course requires the organisation of a field study on a ship. Without the opportunity to carry out an actual field study, the course loses most of its essence. However, it is possible to find alternatives to field study on a ship. This year one student had to be excused and did his field study in a café while traveling. However, we recommend finding study sites that are able to accommodate an entire class to make it possible for teachers to be present in the near vicinity throughout the study.

The feedback given by the ship company that hosted the field study was positive. First, the management of the company has reviewed the observations, analyses and design proposals produced by the students in their reports and found it pertinent and useful. Second, the ship's crew was pleased with the experience and there was no hindering of their work or of the ship operations.

For further development of the course, we are interested in exploring the following directions:

- a course that can be done partially online, allowing the students to design and execute their own field study
- an online-based test of observation skills: photos from previous field studies are shared online and the students are asked to make observations so that we can compare what we have observed with what the students manage to pick-up
- an internal course tailor-made for a company and its own design processes

5. CONCLUSION

From our evaluation, the course seemed to be successful in assisting students in learning a process and techniques to plan and execute a field study. We believe this is a useful contribution to making design education for human-centred design processes available to the maritime industries. This is important because it can affect the safety and efficiency of operations and improve future maritime innovation processes.

6. REFERENCES

1. ABEYSIRIWARDHANE, A. *et al.* (2017) *Human-centred design knowledge into maritime engineering education; theoretical framework.* Australas. J. Eng. Educ. 0, 1–12. doi:10.1080/22054952.2017.1287038
2. ABEYSIRIWARDHANE, A. *et al. (2015). Future ship designers and context of use: setting the stage for human centred design,* Presented at the International Conference on Marine Design, The Royal Institution of Naval Architects.
3. BLOOMBERG,J., Burrell, M., Guest, G. (2009) *An ethnographic approach to design.* Hum.-Comput. Interact. 71–94.
4. BUTTON, G. (2000) *The ethnographic tradition and design.* Des. Stud. 21, 319–332.
5. DEWEY, J. (1938) *Experience and education, First Touchstone Edition* 1997. Simon and Schuster.
6. KOLB, D.A. (1984). Kolb, D.A., 1984. *Experiential learning: Experience as the source of learning and development.* Prentice Hall, Englewood Cliffs, NJ, USA.
7. KUJALA, S. (2003) *User involvement: a review of the benefits and challenges.* Behav. Inf. Technol. 22, 1–16.
8. KUJALA, S., Kauppinen, M., NAKARI, P. and REKOLA, S. (2003) *Field studies in practice: Making it happen*, in: Proceedings of INTERACT 2003. pp. 359–366.
9. KVALE, S. (1996) *Interviews: An introduction to qualitative research interviewing.* Thousand Oaks, California: Sage Publications.
10. KVALE, S. and BRINKMANN, S. (2009). I*nterview: Learning the craft of qualitative research interviewing.* Det Kvalitative Forskningsintervju Oslo Gyldendal Akad.
11. LURAS, S., LUTZHOFT, M., SEVALDSON, B. (2015) *Meeting the complex and unfamiliar: Lessons from design in the offshore industry.* Int. J. Des. 9, 141–154.
12. LURAS, S. and NORDBY, K. (2015) *Shaping designers' sea sense: A guide for design-driven field research at sea.* Presented at the International Conference on Marine Design, The Royal Institution of Naval Architects, London, UK.
13. LURAS, S. and NORDBY, K. (2014) *Field studies informing ship's bridge design.* Presented at the Human Factors in Ship Design and Operation conference, The Royal Institution of Naval Architects, London, UK.
14. LUTZHOFT, M. (2004) *"The technology is great when it works": Maritime Technology and Human Integration on the Ship's Bridge.* Graduate School for Human-Machine Interaction, Division of Quality and Human-Systems, Department of Mechanical Engineering, University of Linköping, Sweden.
15. MALLAM S.C. and LUNDH, M. (2014) *Conceptual ship design, general arrangement & integration of the human element: a proposed framework for the engine department work environment.* Proceedings: 11th International Symposium on Human Factors in Organisational Design and Management & 46th Nordic Ergonomics Society Conference. pp. 829–834.

16. MALLAM, S.C., LUNDH, M. and MACKINNON, S.N. (2015) *Integrating Human Factors Ergonomics in large-scale engineering projects: Investigating a practical approach for ship design.* Int. J. Ind. Ergon. 50, 62–72. doi:10.1016/j.ergon.2015.09.007
17. Moodle - Open-source learning platform | Moodle.org, https://moodle.org/ (Accessed 5.2.17).
18. SALUSTRI, F. and NUEMANN, P. (2017) *Creating a Human-Centric Engineering Design Course.* Presented at the Canadian Engineering Education Association Conference, Toronto, Canada.
19. SCHON, D. (1983). *The reflective practitioner.* Basic Books, New York.
20. TONKINWISE, C. (2005) *The Idealist practice of reflection: Typologies, techniques and ideologies for design researchers.* Presented at the Design Research Society biennal conference, Future Ground, DRS, Melbourne, Australia.